RIGHT TO FOOD
Making it Happen

Progress and Lessons Learned
through Implementation

THE
RIGHT
TO
FOOD

F A O
FIAT PANIS

RIGHT TO FOOD
Making it Happen

Progress and Lessons Learned
through Implementation

The Food and Agriculture
Organization of the United Nations

2011

ISBN 978-92-5-106890-8

Contents

Foreword

Approximately six years have passed since the adoption of the 'Voluntary Guidelines to Support the Progressive Realization of the Right to Adequate Food in the Context of National Food Security' (Right to Food Guidelines). The consensus found at FAO Council in 2004 on these Guidelines represents a milestone in the development of the right to food, reflecting FAO members' vision of a world without hunger, made possible by linking food security instruments to human rights and governance tools tackling the root causes of hunger.

A high number of people persistently suffer from hunger and recurring food crises. In this context, the right to food is an essential element to confront circumstances which have affected so disproportionately the most vulnerable people and deprived them of both access to and the means to procure food. The time has come to bridge the gap between the unacceptable reality of a growing number of starving people and the vision of a world free from hunger.

An approach based on the right to food and good governance can bring about an essential contribution towards bridging this gap. The right to food does not replace existing development efforts towards hunger reduction: it rather brings a new dimension and complements traditional approaches to fight food insecurity. It does this by providing a legal framework and stressing the notions of rights of individuals and obligations of States. It also ensures the establishment of mechanisms to increase accountability of all – citizenry and government as well as development agencies and other stakeholders. By focusing on the most vulnerable, the right to food ensures that targeted action will benefit the hungry without discrimination. By promoting transparency, participation and accountability, it improves the efficiency of measures taken by governments and stakeholders. Finally, by empowering the poor, it ensures that they have a voice, are involved in decision-making and can claim their rights.

In the past few years, FAO has successfully supported several initiatives at country level and has gained considerable experience in promoting the right to food worldwide. This knowledge was brought together during the Right to Food Forum held at FAO headquarters in Rome in October 2008. The Forum provided a platform for over 400 representatives from FAO member countries, civil society organizations, international organizations and academia to share practical experiences and lessons learned from these pilot activities; discuss the progress, constraints and achievements reached so far; and identify new ways to promote the realization of the right to food as a human right. The resulting exchanges contributed to creating a momentum for a strengthened commitment towards promoting the right to food and principles of good governance, particularly at country level.

Recent initiatives have consolidated the right to food both as an objective and as a tool to achieve food security for all. As an objective, the right to food provides an overarching framework that guides efforts at international, national, regional and sub-national levels to address food insecurity and its structural root causes. As a tool, applying human rights principles in policy processes creates a better chance for increased efficiency, effectiveness and impact of policy and operational measures to achieve food security for all. The right to food implies a change of perspective: the hungry cease to be a problem; they become both part of the solution and actors of their own development.

Right to Food – Making it Happen *is the first publication that brings together the practical experiences and lessons learned during the years 2006 to 2009 with the implementation of the right to food at country level, based on the Right to Food Guidelines. It looks at how the right to food has been integrated into policy planning, strategy formulation, programme design and legislative processes throughout Latin America, Africa and Asia. It also highlights successes to be applauded and challenges to be met in five countries that took concrete steps to focus on the right to food when they identified the hungry, conducted assessments, developed strategies, adopted legislation, strengthened coordination and set up monitoring and accountability mechanisms.*

These pages shed light on right to food achievements during the period 2006-2009, as well as progress in advancing its legal, political and institutional dimensions. Such progress paves the way to a more all-encompassing adoption of the right to food and good governance principles at the global and national levels, and also in multilateral agencies dealing with food and agriculture like FAO, guiding their work in the fight against hunger and the attainment of the Millennium Development Goal 1.

May the experiences, conclusions and recommendations in this publication serve as an inspiration for further action and an increased commitment towards a human right that is here to stay.

Jacques Diouf

Director-General
Food and Agriculture Organization of the United Nations

Acknowledgements

This volume represents four years of intensive work by FAO and partners throughout the world in assisting countries in the formulation and implementation of legislation, strategies, policies and programmes recommended in the Voluntary Guidelines to Support the Progressive Realization of the Right to Adequate Food in the Context of National Food Security *(Right to Food Guidelines) approved by FAO Council in 2004. During this time, several right to food initiatives were successfully undertaken and considerable experience gained in advancing the right to food worldwide.*

Many people have contributed to the contents of this document, which is a collection of valuable lessons learned and recommendations reflecting the past years of implementation activities based on the different experiences at country level and assisted by the Guidelines.

FAO is proud to acknowledge the groundbreaking work of a very dedicated team of staff and consultants, led by Barbara Ekwall: Mauricio Rosales, Margret Vidar, Andreas von Brandt, Frank Mischler, Isabella Rae, Julian Thomas, Maartin Immink, Dubravka Bojic Bultrini, Lidija Knuth, Luisa Cruz, Rebecca Kik and Simona Smeraldi. Gabriele Zanolli, Daniela Verona and Tomaso Lezzi were responsible for the layout and presentation of the many quality publications issued during the four-year period and the present publication. The very efficient and affable administrative and secretarial assistance provided by Patricia Taylor, Federico Patimo, Tiziana Tarricone, Sonia Santangelo, and Sophia Mann was also instrumental in ensuring an excellent series of outputs.

The important contributions of our colleagues and partner institutions, as well as the actors in the right to food scene who have provided technical support throughout our important mandate, are gratefully acknowledged. Our thanks go to the many FAO departments that contributed to our endeavours. Their inputs to the different right to food publications, as well as their collaboration with regard to integrating the right to food into FAO's work were of paramount importance in the realization of this work.

FAO wishes to acknowledge the important contribution of institutions such as FIAN International, Germany; and Prosalus, Spain; and is especially grateful for the very fruitful partnership with institutions such as the Technical Secretariat for Food Security and Nutrition (SETSAN) in Mozambique; the Brazilian Action for Nutrition and Human Rights (ABRANDH); the Office of the High Commissioner for Human Rights in Geneva; ESCR-Asia in the Phillippines; the Human Rights Commission in Uganda; the International Budget Partnership in Washington, DC, USA; the University of Manheim in Germany; Akershus University College in Norway; and Makerere University of Uganda. A special word of appreciation is due for the support from and excellent collaboration with the UN Special Rapporteur on the Right to Food, Olivier De Schutter and the former Special Rapporteur, Jean Ziegler.

Achievements in relation to the implementation of the Right to Food Guidelines, aimed at the progressive realization of the right to food, have been remarkable. None of this would have been possible without the support of our donors – in particular Germany, Norway, the Netherlands, Switzerland and Spain.

An important milestone in the generation of information based on concrete country experiences was the Right to Food Forum 2008, held in Rome. FAO is grateful to all those who took part and, in particular, the panellists, speakers and experts, for their presence and for the many important papers and testimonies provided.

Experiences from country implementation were shared by:

Brazil – Rosilene Cristina Rocha, Ministry of Social Development and Fight Against Hunger; Albaneide Peixinho, Ministry of Education; and Elisabetta Recine, National Food and Nutrition Security Council.

Guatemala – Lorena Pereira, Presidential Commission on Human Rights; Luis Enrique Monterroso, FAO Right to Food Project.

India – Colin Gonsalves, Human Rights Law Network; Biraj Patnaik, Office of the Commissioners to the Indian Supreme Court on the Right to Food Case; and Aruna Sharma, National Human Rights Commission.

Mozambique – Francisca Cabral, Technical Secretariat for Food Security and Nutrition (SETSAN); Aida Libombo, Ministry of Health; and Ali Bachir Macassar, Ministry of Justice.

Philippines – Leila de Lima, Human Rights Commission; Dolores B. de Quiros Castillo, National Anti-Poverty Commission; Cecilia Florencio, NGO Liaison Committee for Food Security and Fair Trade (PNLC).

Uganda – David O.O. Obong, Ministry of Water and Environment; Aliro Omara, Uganda Human Rights Commission; and Gerald Tushabe, Human Rights Network, Hurinet.

FAO wishes to give special thanks to the following persons for their contributions as facilitators of the Forum's five panels: Marcos Arana-Cedeño, National Institute of Nutrition, Mexico; Vincent Calderhead, Nova Scotia Legal Aid, Canada; Charlotte McClain-Nhlapo, World Bank, USA (formerly with the South African Human Rights Commission); Sandra Ratjen, FIAN International, Germany; and Michael Windfuhr, Bread for the World, Germany.

Lively and fruitful discussions emerged thanks to the following panellists:
Wenche Barth Eide, University of Oslo, Norway; Anne C. Bellows, Hohenheim University, Germany; Asbjorn Eide, University of Oslo, Norway; Christophe Golay, The Graduate Institute of International Development Studies (IHEID), Switzerland; Colin Gonsalves, Human Rights Law Network, India; Asako Hattori, Office of the High Commissioner for Human Rights (OHCHR); Julio Cesar Guerrero Arraya, National Council for Food and Nutrition (CONAN), Bolivia; Leila de Lima, Human Rights Commission, Philippines; Aida Libombo, Ministry of Health, Mozambique; Guido Lombardi, Presidential Commission on the Millennium Development Goals and Sustainable Development, Peru; Duciran van Marsen Farena, Federal Public Prosecutor, Brazil; Carlota Merchán, Prosalus, Spain; Khalid S. Mohammed, Ministry of Agriculture, Livestock and Environment, Zanzibar; Arne Oshaug, Akershus University, Norway; Lorena Pereira, Presidential Commission on Human Rights, Guatemala; Dolores de Quiros Castillo, National Anti-Poverty Commission, Philippines; Elisabetta Recine, Brazilian Action for Nutrition and Human Rights (ABRANDH), Brazil; Eibe Riedel, Committee on Economic, Social and Cultural Rights, Switzerland; Myrna Romero Arana, National Council for Food and Nutrition (CONAN), Bolivia; A. Byaruhanga Rukoko, Makerere University, Uganda; Carole Samdup, Rights and Democracy, Canada; Francisco Sarmento, ActionAid International, Brazil; Flavio Valente, FIAN International, Germany; Martin Wolpold-Bosien, FIAN International, Germany; and Dora Zeledon, National Assembly, Nicaragua.

A special word of appreciation goes to Marc Cohen, Forum Rapporteur, who provided an in-depth report of the meeting which is included in the Annex to this document.

The detailed case studies contained in Part THREE indicate how selected countries have included the right to food in their assessment, strategy development, legislation and coordination procedures, in keeping with FAO's 'seven-step process'. These studies are based on papers developed for the Forum. The authors of these papers are acknowledged as follows: Elisabetta Recine and Frank Mischler for Brazil; Mauricio Rosales and Luis Enrique Monterosso for Guatemala; Aruna Sharma and Margret Vidar for India; Lazaro dos Santos and Cecilia Luna for Mozambique; and Peter Rukundo for Uganda.

Right to Food – Making it Happen *was developed under the overall guidance of Barbara Ekwall, the Coordinator of the Right to Food Team, along with contributions from Margret Vidar, Mauricio Rosales, Frank Mischler and Isabella Rae. Maartin Immink, Rebecca Kik and Sherry Ajemian brought together the different contributions linking the threads to bring coherence and structure to the text which was edited by Barbara Rae.*

Acronyms and Abbreviations

ABRANDH	Ação Brasileira para Nutrição e Direitos Humanos (Brazilian Action for Nutrition and Human Rights)
AJEHR	African Journal on Ethics and Human Rights
AIDS	Acquired Immune Deficiency Syndrome
APL	Above poverty line (India)
BPL	Below poverty line (India)
CESCR	UN Committee on Economic, Social and Cultural Rights
CICSAN	Centro de Információn y Coordinación de Seguridad Alimentaria y Nutricional (Food Security and Nutrition Information and Coordination Centre – Guatemala)
CIIDH	Centro Internacional de Investigación en Derechos Humanos (International Centre for Human Rights Research)
CONAN	Consejo Nacional de Alimentacíon y Nutrición (National Council for Food and Nutrition Security – Bolivia)
CONSAN	Conselho Nacional de Segurança Alimentar e Nutricional (National Council for Food and Nutrition Security – Mozambique)
CONASAN	Consejo Nacional de Seguridad Alimentaria y Nutricional (National Food and Nutrition Security Council – Guatemala)
CONASSAN	Comisión Nacional de Soberanía y Seguridad Alimentaria y Nutricional (National Commision for Food and Nutrition Security and Sovereignty)
CONSEA	Conselho Nacional de Segurança Alimentar e Nutricional (National Council on Food and Nutrition Security – Brazil)
COMAN	Consejo Municipal de Alimentacíon y Nutrición (Municipal Council for Food and Nutrition – Bolivia)
COPREDEH	Comisión Presidencial Coordinadora de la Política del Ejecutivo en Materia de Derechos Humanos (Presidential Commission on Human Rights – Guatemala)
CPI	Consumer price index
CPR	Civil and political rights
CSO	Civil society organizations
DDPR	Department of Disaster Preparedness and Refugees (Uganda)
DES	Dietary energy supply
ECA	Essential Commodities Act (India)
ECLAC/CEPAL	Economic Commission for Latin America and the Caribbean (Comisión Económica para América Latina y el Caribe)

ESAN	Estratégia de Segurança Alimentar e Nutricional (Food Security and Nutrition Strategy – Mozambique)
ESCR	Economic, social, and cultural rights
FAO	Food and Agriculture Organization of the United Nations
FIAN	FoodFirst Information and Action Network
FSN	Food security and nutrition
GDP	Gross domestic product
GIA	Grupo de Instituciones de Apoyo (Institutional Support Group – Guatemala)
GIISAN	Grupo Inter-Institucional de Seguridad Alimentaria y Nutricional (Inter-Institutional Group for Food and Nutritional Security – Guatemala)
HIPC	Highly Indebted Poor Countries
HIV	Human immunodeficiency virus
HSSP	Health Sector Strategic Plans (Uganda)
IBGE	Instituto Brasileiro de Geografia e Estatística (Brazilian Institute of Geography and Statistics)
IBSA	Indicators, Benchmarks, Scoping and Assessment
ICDS	Integrated Child Development Scheme (India)
ICESCR	International Covenant on Economic, Social and Cultural Rights
IDP	Internally displaced persons
IFAD	International Fund for Agricultural Development
INCOPAS	Instancia de Consulta y Participación Social (Social Participation and Consultation Authority – Guatemala)
INR	Indian Rupees (India)
IPRFD	International Project on the Right to Food in Development
JLOS	Justice Law and Order Sector (Uganda)
LAP	Legal Aid Project (Uganda)
LOSAN	Lei Orgânica de Segurança Alimentar e Nutricional (Federal Law on Food and Nutrition Security – Brazil)
LRA	Lords Resistance Army (Uganda)
LRC	Law Reform Commission (Uganda)
MAAIF	Ministry of Agriculture, Animal Industry and Fisheries (Uganda)
MDGs	Millennium Development Goals
MDS	Ministry of Social Development and Fight Against Hunger (Brazil)
MFPED	Ministry of Finance, Planning, and Economic Development (Uganda)

MOH	Ministry of Health (Uganda)
MTCS	Medium Term Competitiveness Strategy (Uganda)
NAADS	National Agricultural Advisory Services (Uganda)
NGO	Non-governmental organization
NHRC	National Human Rights Commission (India)
NREGA	National Rural Employment Guarantee Act (India)
OHCHR	Office of the High Commissioner for Human Rights
PANTHER	Participation, Accountability, Non-discrimination, Transparency, Human Dignity, Empowerment and Rule of Law
PARPA	Plano de Acção para a Redução da Pobreza Absoluta (Plan of Action for a Reduction in Absolute Poverty – Mozambique)
PDS	Public Distribution System (India)
PEAP	Poverty Eradication Action Plan (Uganda)
PMA	Plan for the Modernization of Agriculture (Uganda)
PPP	Private Public Partnership (Uganda)
PROCADA	Proyecto de Promoción y Capacitación para la Implementación Progresiva del Derecho a la Alimentación en Guatemala (Promotion and Training Project for the Progressive Realization of the Right to Food in Guatemala)
PRSP	Poverty Reduction Strategy Paper
PUCL	People's Union for Civil Liberties (Rajasthan – India)
ROSA	Rede de Organizações para a Soberania Alimentar (Network of Food Sovereignty Organizations – Mozambique)
SESAN	Secretaría de Seguridad Alimentaria y Nutricional (Secretariat for Food and Nutrition Security – Guatemala)
SETSAN	Secretaria Técnica de Segurança Alimentar e Nutricional (Technical Secretariat for Food Security and Nutrition – Mozambique)
SINASAN	Sistema Nacional de Seguridad Alimentaria y Nutricional (National Food and Nutrition Security System – Guatemala)
SINASSAN	Sistema Nacional de Soberanía y Seguridad Alimentaria y Nutricional (National System for Food and Nutrition Security and Sovereignty)
SISAN	Sistema Nacional de Segurança Alimentar e Nutricional (National Food and Nutrition Security System – Brazil)
UBOS	Uganda Bureau of Statistics
UDHS	Uganda Demographic and Health Survey
UEM	Universidade Eduardo Mondlane (Eduardo Mondlane University – Mozambique)

UFNC	Uganda Food and Nutrition Council
UFNP	Uganda Food and Nutrition Policy
UFNSIP	Uganda Food and Nutrition Strategy and Investment Plan
UHRC	Uganda Human Rights Commission
ULS	Uganda Law Society
UN	United Nations
UNBS	Uganda National Bureau of Standards
UNDP	United Nations Development Programme
UNHCR	United Nations High Commissioner for Refugees
UNICEF	United Nations Children's Fund
USAID	US Agency for International Development
VAT	Value Added Tax
WFP	World Food Programme
WHO	World Health Organization
WTO	World Trade Organization

Introduction

The right to food is an integral part of a vision of a world without hunger, where every child, woman and man can feed himself or herself in dignity. It is a human right formally recognized by the great majority of States. While there is consensus about the vision, States have been slow in putting this human right into practice. And yet, the right to food is far from being a slogan or an academic theory of development. It is about concrete actions and practical solutions. These cover several domains and, in addition to governments, involve key actors ranging from individuals to non-governmental organizations, academia, media, UN human rights institutions, and the private sector.

Since the adoption of the *Voluntary Guidelines to support the progressive realization of the right to adequate food in the context of national food security* (Right to Food Guidelines) by FAO Council in 2004, a number of countries, associations, individuals and organizations embarked on putting the right to food into practice through advocacy, policy making, legislation, monitoring, assessment and the strengthening of institutions. These insights, experiences and lessons learned were presented and discussed at the Right to Food Forum held at FAO from 1st to 3rd October 2008. The case studies from five different countries were discussed in more detail on that occasion.

The publication *Right to Food – Making it Happen* is a summary of three days of exchange on different issues related to country level implementation of the right to food that took place at FAO's Right to Food Forum, including a more in-depth discussion on five countries. It is an effort to share real and practical experiences in a human rights-based approach to combat food insecurity with a particular focus on one fundamental human right – the right to food – as recognized by the Universal Declaration of Human Rights and the International Covenant on Economic, Social and Cultural Rights (ICESCR).

The relevance of promoting right to food as a strategy to fight hunger, as opposed to simply promoting provision of food and development aid, is clearly reflected in FAO's mandate and strategic objectives. Through its efforts to support the formulation of better policy and strategy options on a global scale, FAO plays a key role on the common agenda. It aims to reduce hunger by creating platforms for sharing technical expertise, strategic policy alternatives, and outcomes of implementation of food security strategies adopted by member nations. The Right to Food Forum represented one of these platforms.

This publication is intended to disseminate the lessons learned from discussions that occurred during the event. The overall theme was right to food as a strategy promoted within a human rights-based approach and implemented throughout development assistance programs. It thus links right to food to the overall objective achieving global food security.

The target audience for this publication are development specialists that provide food security policy advice to member nations; UN and non-governmental agencies promoting human rights; official policymakers and legislators of donor as well as developing countries; non-profit organizations involved in food aid and food assistance; and finally all stakeholders involved in country level projects and programmes with the aim of reducing hunger. The intent of the publication is to show that responding to hunger and food insecurity requires coordination of national food security initiatives and increased policy coherence. One must also stress the importance of strengthening the actual institutions, mechanisms, partners and sectors that promote or support the right to food.

Right to Food - Making it Happen comprises three parts. Part ONE clarifies concepts related to the human right to adequate food, shows how it can strengthen efforts to reduce world hunger, describes implementation steps and FAO's work in this area. Part TWO offers a synthesis of the Right to Food Forum. It reflects the rich deliberations, outputs and lessons learned from panels that dealt with advocacy and training, legislation, targeting and assessment, monitoring, strategy and coordination, supplemented by best practice examples in different countries worldwide. Part THREE provides five country case studies, describing the lessons learned and indicating the way forward to a progressive realization of the right to food in Brazil, Guatemala, Mozambique, India and Uganda. The Appendix to this publication contains the full texts of the Opening speeches and the report by the Forum Rapporteur.

Part ONE

BRIEF INTRODUCTION TO THE RIGHT ON FOOD AS A HUMAN RIGHT

VJ Villafranca / IRIN

I. THE RIGHT TO FOOD

"Striving to ensure that every child, woman and man enjoys adequate food on a regular basis is not only a moral imperative and an investment with enormous economic returns; it also signifies the realization of a basic human right."

(Jacques Diouf, Director-General, FAO)[1]

The right to adequate food[2] as a basic human right was first recognized in the Universal Declaration of Human Rights in 1948, as part of the right to a decent standard of living (Art. 25):

'Everyone has the right to a standard of living adequate for the health and well being of himself and his family, including food...' It became legally binding when the International Covenant on Economic, Social and Cultural Rights (ICESCR) entered into force in 1976. Since then, many international agreements have affirmed the right to food, among them the Convention on the Elimination of All Forms of Discrimination against Women (CEDAW), 1979, and the Convention on the Rights of the Child (CRC), 1989.

To date, 160 states have ratified the ICESCR and are legally bound by its provisions. In Article 11, the ICESCR stipulates that the States Parties *'recognize the right of everyone to an adequate standard of living for himself and his family, including adequate food'* and affirms the existence of *'the fundamental right of everyone to be free from hunger.'*

Freedom from hunger is considered to be the minimum level that should be secured for all, independent of the level of development of a given State. But the right to food not only implies being free from hunger: the Committee on Economic, Social and Cultural Rights (CESCR) defined the right to food in General Comment No. 12 as follows: *'The right to adequate food is realized when every man, woman and child, alone or in community with others, has physical and economic access at all times to adequate food or means for its procurement.'*[3] Furthermore, the Committee stresses that the right to adequate food *'shall not be interpreted in a narrow and restrictive sense which equates it with a minimum package of calories, proteins and other specific nutrients.'*[4] Thus, important elements of food practices, education on hygiene, training on nutrition, and concerns such as provision of health, care and breastfeeding can enter the realm of discussion on right to food policy design, implementation and monitoring.

This means that each person has the right to have access to the resources necessary to produce, earn and purchase adequate food, not only to prevent hunger but also to ensure good health and well-being. General Comment No. 12 considers that the core content of the right to adequate food implies two elements. The first is *'the availability of food in a quantity and quality sufficient to satisfy the dietary needs of individuals, free from adverse substances, and acceptable within*

---

1   FAO. 2005. Jacques Diouf in Foreword to the *Voluntary Guidelines to Support the Progressive Realization of the Right to Adequate Food in the Context of National Food Security*, p. iv. Rome.

2   This publication frequently uses the shortened form 'Right to Food' meaning, in all cases, the human right to adequate food as enshrined in the ICESCR, Art. 11, and described in *General Comment 12* of the CESCR.

3   CESCR. *General Comment 12, The right to adequate food*. E/C.12/1999/5, par. 6.

4   *Ibidem*, par. 6.

a given culture'[5] and the second element is *'the accessibility of such food in ways that are sustainable and that do not interfere with the enjoyment of other human rights.'*[6]

The international human rights instruments place the primary responsibility for the realization of the right to food on the State, and identify three categories of State obligations: the obligations to respect, protect and fulfil (facilitate and provide). As stated in General Comment 12, *'the obligation to respect existing access to adequate food requires State parties not to take any measures that result in preventing such access.'*[7]

'The obligation to protect requires measures by the State to ensure that enterprises or individuals do not deprive individuals of their access to adequate food. The obligation to fulfil (facilitate) means that the State must pro-actively engage in activities intended to strengthen people's access to and utilization of resources and means to ensure their livelihood, including food security.'[8] This means that the State has to create a legal, policy and institutional environment that enables people to access safe and nutritious food in ways that fully respect human dignity – either through procurement or production. Fulfil (provide) means that *'whenever an individual or group is unable, for reasons beyond their control, to enjoy the right to adequate food by means at their disposal, States have the obligation to fulfil (provide) the right directly,'*[9] an example of which could be supplying food aid.

Furthermore, the States Parties to the ICESCR are obliged to take steps, individually and through international assistance and cooperation, especially economic and technical, to the maximum of their available resources, with a view to achieving progressively the full realization of the right to food.[10] The progressive realization of the right to food implies that full realization of this human right requires time and, therefore, cannot be achieved immediately. It also implies the principle of non-retrogession in implementing human rights whereby once a commitment is made to protect a human right such as the right to food it can not subsequently be withdrawn. Therefore, the standard of protection of a human right in effect can not be lowered and must be progressively and actually realized. However, as a minimum immediate core obligation, States must ensure freedom from hunger in their territory.

The international community confirmed its political will and the need to fully respect, protect and fulfil the right to food on several occasions like the World Food Summits of 1996 and 2002. During the World Food Summit 2002 the idea emerged of developing a voluntary instrument on the right to food; as a result, that same year the FAO Council created the Intergovernmental Working Group (IGWG) on the right to food, to design, discuss and negotiate this voluntary instrument. The outcome of that process led to the *Voluntary Guidelines to Support the Progressive Realization of the Right to Adequate Food in the Context of National Food Security* (**Right to Food Guidelines**), which were adopted by consensus at the FAO Council meeting

---

5   CESCR. *General Comment 12, The right to adequate food*. E/C.12/1999/5, par. 8.

6   *Ibidem*, par. 8.

7   CESCR. *General Comment 12, The right to adequate food*. E/C.12/1999/5, par. 15.

8   *Ibidem*, par. 15.

9   *Ibidem*, par. 15.

10 CESCR. *General Comment 12, The right to adequate food*. E/C.12/1999/5, par. 36.

in 2004. States are encouraged to apply these Guidelines when developing their strategies, policies, programmes and activities, and to do so with no discrimination of any kind.

The experience gained during the past few years shows that the Right to Food Guidelines are a valuable instrument for helping States to promote the right to food. Although States showed their political commitment in the year 2000 to halve the proportion of people who suffer from hunger by 2015 as agreed in the Millennium Summit, the number of hungry and undernourished people rose to more than 1. 02 billion people worldwide in 2009.[11] This shows the still existing gap between the standards set in international treaties and the prevailing situation in many parts of the world. Considering the continuing existence of and current increase in food insecurity, implementation of the Guidelines is more pertinent today than ever before.

11   http://www.fao.org/news/story/en/item/20568/icode

II. FOOD SECURITY AND THE RIGHT TO FOOD

"Food security exists when all people, at all times, have physical and economic access to sufficient, safe and nutritious food to meet their dietary needs and food preferences for an active and healthy life."

(World Food Summit Plan of Action – 1996)[12]

The Right to Food Guidelines reflect the consensus among FAO member countries on what needs to be done in all critically relevant policy areas in order to promote food security through a human rights based approach. Governments agreed on the full meaning of the right to food, on what it entails in practice and what needs to be done in areas such as food aid, nutrition, education strategies, access to resources, and legal frameworks and institutions, in order to realize this right.

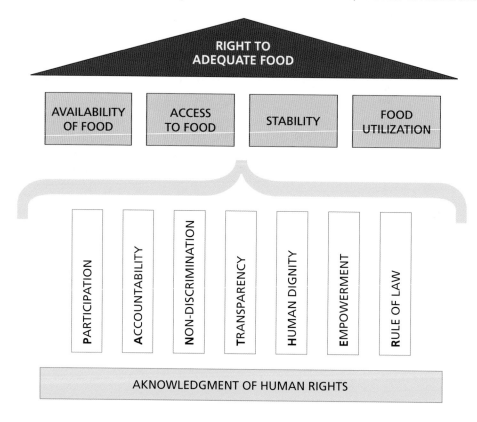

The right to food strengthens **the four pillars of food security** – availability, access, stability of supply, and utilization – with human rights principles. The areas of action outlined in the Right to Food Guidelines are fully consistent with the 'twin-track' approach to food security as developed by FAO, jointly with IFAD and WFP. Track One aims at creating opportunities for the food insecure and vulnerable to improve their livelihoods by promoting development, particularly agricultural and rural development, through policy reforms and investments in agriculture and related sectors.

---

12   FAO. 1996. *Rome Declaration on World Food Security and World Food Summit Plan of Action*. Rome.
     (Available at http://www.fao.org/docrep/003/w3613e/w3613e00.HTM).

Track Two involves direct action to fight hunger through food and non-food programmes that provide the hungry with immediate access to food. Through its human rights based approach to food security, the right to food puts *the people* at the centre of development, so that they are recognized as right holders and not as mere beneficiaries.

The right to food offers a coherent framework within which to address critical **governance** dimensions in the fight against hunger and malnutrition. Whereas many food security policies and programmes deal with essential technical issues, still both governance and human rights have to be addressed in order to ensure the effectiveness and sustainability of food security work.

The right to food provides a voice to a wide array of relevant stakeholders; moreover, it establishes the seven 'PANTHER' principles that should govern decision-making and implementation processes: **P**articipation, **A**ccountability, **N**on-discrimination, **T**ransparency, **H**uman dignity, **E**mpowerment and the **R**ule of law (the first letters of each together forming the acronym).

Originating from different human rights treaties, these principles relate to the process that should be followed in addressing the right to adequate food.

This right to food approach contributes to strengthening relevant public institutions and coordination mechanisms with regard to implementation. It integrates partners such as civil society organizations, human rights commissions, parliamentarians and government sectors, besides those dealing with agriculture, and provides further justification for investment in hunger reduction. In addition, the right to adequate food provides a legal framework, the concept of rights and obligations, and the relevant mechanisms needed to achieve **accountability** and to promote the **rule of law**.

As to the content of food security work, the right to adequate food concept introduces additional – mainly legal – instruments that ensure access by the most vulnerable people to, among other things, income earning opportunities and social protection in particular. It uses the power of law to strengthen the means of implementation. It also enhances governmental action by introducing administrative, quasi-judicial and judicial mechanisms to provide effective remedies, by clarifying the rights and obligations of right holders and duty bearers, and by strengthening the human rights mandate of relevant institutions.

III. THE RIGHT TO FOOD IN FAO

"...the international community, and the UN system, including FAO, as well as other relevant agencies and bodies according to their mandates, are urged to take actions in supporting national development efforts for the progressive realization of the right to adequate food in the context of national food security."

(Right to Food Guidelines – Part III: International Measures, Actions and Commitments)

The right to food requires political commitment at the highest level. Heads of State and Governments participating in the World Food Summit of 1996 reaffirmed *'the right of everyone to have access to safe and nutritious food, consistent with the right to adequate food and the fundamental right of everyone to be free from hunger.'*[13]

The right to food has been part of FAO's mandate since its inception and is firmly embedded within FAO's Strategic Framework adopted in 2009. In fact, it is a key component of the Organizational Result H2 that contributes to achieving FAO's Strategic Objective H – Improved Food Security and Better Nutrition.

An important milestone towards the practical implementation of the right to food was achieved with the adoption of the Right to Food Guidelines by the FAO Council in 2004. The Council also asked FAO to support interested member countries in putting the Guidelines in practice. This led to the establishment of the Right to Food Unit at FAO. The main functions of the Unit are to raise awareness of right to food, to develop tools and mechanisms for implementation and to provide technical expertise and policy advice to countries in their efforts to formulate and implement legislation, strategies, policies and programmes based on the Guidelines. Its mandate also includes facilitating the integration of right to food principles and approaches in FAO's normative and technical assistance work.

The work of the Right to Food Team is broken down into the following strategic areas:[14]

- **Advocacy and training**: to promote awareness and understanding of the right to food among right holders and duty bearers; to increase right holders' capacity to demand accountability of public actions and claim their rights; and to increase the capacity of duty bearers to fulfil their obligations and responsibilities;

- **Legislation and accountability**: to promote and facilitate the incorporation of the right to food into national constitutions and laws; to ensure conformity of domestic legislation with State obligations contained in international human rights instruments related to the right to food; and to assist countries in the implementation of adequate recourse mechanisms;

- **Information and assessment**: to assess existing legal, institutional and policy frameworks for the purpose of finding more conducive ways to implement measures that promote the right to food; to identify the hungry and the root causes for their food insecurity; and to assist in the analysis of food security and nutrition vulnerability with a view to developing targeted policies and programmes for the most needy;

---

13   FAO. 1996. *Rome Declaration on World Food Security and World Food Summit Plan of Action*. Rome. (Available at http://www.fao.org/wfs/index_en.htm).

14   FAO. 2006. *The Right to Food in Practice: Implementation at the National Level*. Rome.

- **Benchmarks and monitoring**: to develop and implement monitoring systems that focus on progress in the realization of the right to food; to analyze the positive or negative impacts of policies and programmes on achieving the right to food; and to examine whether public action is implemented in accordance with human rights principles.

- **Strategy and coordination**: to develop and implement policies and programmes as part of a national strategy to achieve the right to food for all; and to ensure that such policies and programmes are properly coordinated and involve all relevant sectors while promoting broad-based participation by civil society and grass roots organizations.

During the past four years, the Right to Food Team has also facilitated interaction between different actors and is engaged in documenting and analyzing much of the experience of governments, NGOs and academic institutions throughout the world. Furthermore, as human rights are at the core of the UN mandate, work on the right to food provides additional entry points for strengthening collaboration between FAO and the UN system.

Furthermore, in order to give strong support for the implementation of the Right to Food Guidelines at country level, the Right to Food Team has developed a **Methodological Toolbox on the Right to Food** which was launched in October 2009. This Toolbox comprises a series of analytical, educational and normative tools in five manuals, providing guidance and hands-on advice to countries on the practical aspects of the implementation of the right to adequate food. It is also an essential contribution to strengthening in-country capacity to implement this human right. The Methodological Toolbox on the Right to Food contains the following manuals:

1. **Guide on Legislating for the Right to Food**

2. **Methods to Monitor the Human Right to Adequate Food** (Volume I/ II)

3. **Guide to Conducting a Right to Food Assessment**

4. **Right to Food Curriculum Outline**

5. **Budget Work to Advance the Right to Food**.

The first manual, **Guide on Legislating for the Right to Food**, provides assistance to legislators and lawyers on how to integrate the right to food into different levels of national legislation. It describes various ways of protecting the right to food in State constitutions, provides step by step guidance on drafting a framework law and a methodology for reviewing the compatibility of sectoral laws with the right to food. Several country examples and experiences are included. This manual also includes a CD of a Legal Database that contains the full text of relevant national legislation on right to food. The second manual, **Methods to Monitor the Human Right to Adequate Food**, provides different methodologies for monitoring the right to adequate food. It is addressed to technical staff in public sector institutions and civil society organizations responsible for planning and monitoring food security, nutrition and poverty reduction policies and programmes. The third manual, **Guide to Conducting a Right to Food Assessment**, provides assistance to governments, civil society and other stakeholders in the assessment of the right to food situation at national level. The fourth manual, **Right to Food Curriculum Outline**, provides a unique basis for education, training and advocacy on the right to food. It aims to contribute to strengthening in-country capacity to implement this human right and can be used as a reference guide for university lecturers, technical assistance experts, instructors and trainers in developing special courses or full training programmes on the right to food. The fifth manual, **Budget Work to Advance the Right to Food**, is a valuable tool for civil society, human rights defenders, interested legislators and government institutions, exploring the complex ways that government budgets relate to the realization of the right to food. It gives a 10-step guide to the process of building a right to food case and also examines three related case studies.

This Toolbox is available in digital version on FAO's Right to Food website at: http://www.fao.org/righttofood. This site also comprises the aforementioned Legal Database as well as a Virtual Library, numerous publications, information items, tools, reports and e-learning materials.

IV. IMPLEMENTATION OF THE RIGHT TO FOOD

"Developed and developing countries should act in partnership to support their efforts to achieve the progressive realization of the right to adequate food."

(Right to Food Guidelines – Part III: International Measures, Actions and Commitments)

Growing numbers of people worldwide are demanding action on the right to food; many governments are heeding this call and taking initiatives to strengthen the implementation of this right.

In spite of significant efforts to improve the food and nutrition situation in many countries, the number of hungry people in many parts of the world has risen over the last decade. This publication comes at a time when decision makers in FAO member countries, and other stakeholders, are concerned about the effects of climate change, bio-fuel production and sharply rising food prices on the food security of large segments of the world's population.

Hunger is a human rights issue. Never before has this link been recognized so strongly as in the context of the global food crisis that began in 2007 with soaring food and energy prices, and continues today as a result of the financial crisis and economic slowdown. FAO estimated that by the end of 2008 the number of undernourished people worldwide had increased to 963 million[15], while in 2009 the figure soared to reach 1.02 billion[16].

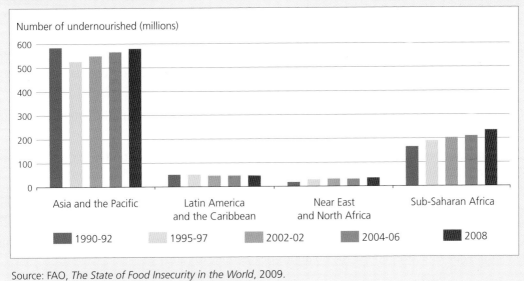

Undernourishment on the rise throughout the world

Number of undernourished in selected regions, 1990-92 to 2008

Number of undernourished (millions)

Legend: 1990-92, 1995-97, 2002-02, 2004-06, 2008

Regions: Asia and the Pacific, Latin America and the Caribbean, Near East and North Africa, Sub-Saharan Africa

Source: FAO, *The State of Food Insecurity in the World*, 2009.

---

15   http://www.fao.org/news/story/en/item/8836/icode
16   http://www.fao.org/publications/sofi/en

The poor allocate a large share of their household resources to food purchases. Therefore, in times of rising food costs they become even more vulnerable to the risk of hunger and malnourishment. Achievement of Millennium Development Goal 1, that is, to halve the number of hungry people in the world by 2015, has now become an even greater challenge.

Clearly, it is unrealistic to expect a country to eliminate hunger and malnutrition overnight. The progressive realization of the right to food means that the number of hungry and malnourished people will diminish over time. Most of the action needed in order to implement the right to food takes place at national level. Consequently, countries should allocate as much as possible of their available resources towards achieving the right to food so as to ensure that, within a reasonable period of time, the number of hungry people in their midst will diminish – both effectively and sustainably. Political will and adherence to international standards are the keys to achieve this goal. At national level, the right to food is being increasingly integrated into constitutions and legislation, as well as policies, strategies and programmes. Furthermore, several court cases around the world were upheld and resulted in the enforcement of the right to food – or some aspects of it – as in the examples of India and Brazil.

The implementation and promotion of the right to food at country level is generally conceived as a seven-step process.

The seven steps are:

1. Identification and characterization of the hungry and food-insecure for targeted policies and programmes;

2. Assessment of the legal, policy and institutional environment, and of current budgetary allocations and expenditures, to indicate the need for food security policy change and other measures;

3. Development of human rights based food security strategies with verifiable and time-bound objectives and targets, institutional responsibilities, mechanisms for coordinated action, and adequate monitoring systems;

4. Definition of inter-institutional coordination mechanisms, and participation by non-governmental sectors;

5. Integration of the right to food in national legislation, such as a constitution or a framework law, to establish long-term binding obligations for State actors and other stakeholders;

6. Establishment of a monitoring system to assess the implementation process and the impact of policies and programmes with the eventual goal of holding governments accountable; and

7. Establishment of adequate judicial, quasi-judicial and/or administrative recourse mechanisms that provide effective redress for violations of the right to food.

STEPS TO REALIZE THE RIGHT TO FOOD

1. IDENTIFY THE HUNGRY AND POOR
2. CONDUCT A THOROUGH ASSESSMENT
3. ELABORATE A SOUND FOOD SECURITY STRATEGY
4. ALLOCATE OBLIGATIONS AND RESPONSIBILITIES
5. CREATE A LEGAL FRAMEWORK
6. MONITOR PROGRESS
7. ENSURE RESOURCE MECHANISMS

CAPACITY
BUILDING

The Right to Food Team, together with many other units in FAO, is ready and available to provide assistance to member countries in different parts of the world regarding each of the above-mentioned steps. Its partners, such as FIAN International, ActionAid, Prosalus, Right to Food India, ABRANDH and ESCR-Asia, are engaged in right to food campaigns to empower people to claim their rights. Information is key for right holders to be able to claim their rights and for governments to abide by their human rights obligations. Therefore, capacity development is an integral part of taking action at the country level with the seven steps mentioned above.

V. THE TIME TO ACT IS NOW

"We must build on what was done last year, sustain our successes and scale up our responses, especially as the financial crisis compounds the impact of the food crisis. We must continue to meet urgent hunger and humanitarian needs by providing food and nutrition assistance and safety nets, while focusing on improving food production and smallholder agriculture. This is the 'twin-track' approach taken in the Comprehensive Framework for Action. We should be ready to add a third track – the right to food – as a basis for analysis, action and accountability".[17]

(UN Secretary General Ban Ki-Moon)

Much has been achieved towards the realization of the right to food in the last decade but there is still a lot more to be done. Effective action to implement this right and reverse current food insecurity trends is not only a moral imperative, but it also makes good economic sense: reduced food insecurity and malnutrition contribute to accelerating poverty reduction and enhancing socio-economic development.

It is of the utmost importance to move quickly in implementing the right to food in policies, strategies and laws necessary to ensure the progressive and full realization of this fundamental right. For more than one billion people in the world today, the human right to adequate food is not being met. These people should not have to wait any longer.

---

17   UN Secretary General Ban Ki-Moon. 26-27 January 2009. Closing ceremony speech at "High Level Meeting on Food Security for All", hosted by the Government of Spain. Madrid.

Gabriele Zanolli / FAO

Part TWO

REPORT ON THE RIGHT TO FOOD FORUM 2008

I. INTRODUCTION

FAO hosted the Right to Food Forum from the 1st to the 3rd of October 2008 at its headquarters in Rome. The purpose of the Forum was to extract lessons learned from three years of experience at country level, and to formulate practical recommendations for moving ahead with the progressive realization of the right to food. For the first time at international level, FAO member countries, practitioners and stakeholders, came together to share their experiences in implementing the right to food, learn from experiences in other countries, and discuss ways of promoting and accelerating the implementation of the *Voluntary Guidelines to support the progressive realization of the right to adequate food in the context of national food security* (Right to Food Guidelines).

The Right to Food Forum included a number of plenary sessions during which practitioners and international experts provided an overview of the current global challenges and the actual status of the implementation of the right to food in Brazil, Guatemala, India, Mozambique and Uganda.[18] Subsequently, parallel panel sessions brought together international experts from governments, universities, international organizations and civil society to discuss lessons learned on advancing the right to food in different areas of action. In addition, several side events were held to provide an opportunity for all stakeholders to present case studies and illustrate the many interventions at global and national level. A closing session took place during which conclusions were drawn and agreement on a way forward was identified to strengthen efforts in view of the realization of the right to adequate food for all.

In this part, the reader is invited to reflect on the right to food as a real policy option in the design of food security strategies. He or she is called to consider the key challenges when implementing this approach to development aid and assistance. The lessons learned with the implementation of the right to food on the ground serve as a roadmap for those countries considering a human rights-based approach to fighting hunger. Development frameworks and strategies have much to gain from adopting the right to food as a starting point to improve the governance of food systems. It adds value to food security interventions by focusing on participation and accountability and strengthens the entitlement of the poor and hungry to a life of dignity. It goes without saying that when the commitment at country level is high, and it is not only top-down but also bottom up, the results are tangible.

---

18 The country case studies, which served as a basis for the discussions, have been adapted and updated and are included in Part THREE of the present publication.

II. SYNTHESIS OF THE PANEL SESSIONS

The following is a synthesis of panel discussions held during the Right to Food Forum from 1 October 2008 to 3 October 2008. It reflects the fruitful dialogues, outputs and lessons learned which stem out of the panels. The Forum conclusions and recommendations are intended as a guide to the way forward in implementing the right to food in the future.

Experts from international organizations, governments, civil society and universities participated in the sessions comprising five separate panels, reflecting the five thematic working areas for putting the right to food into practice: (a) Strong Voices: Advocacy and Training, (b) Accessible Justice: Legislation and Accountability, (c) Right Targets: Information and Assessment, (d) Durable Impact: Benchmarks and Monitoring, and (e) Effective Action: Strategy and Coordination.

The synthesis of each of the panel's discussions indicates the main issues dealt with, including the relevant country experiences and challenges met. The lessons learned and conclusions related to the concrete experiences with the implementation of the right to food are summarized together with a set of forward-looking recommendations. Some specific country experiences are highlighted, indicating the way in which challenges were addressed, what specific action was undertaken and with what result. The country experiences described were supplemented by documentation developed by FAO throughout the implementation process of its right to food projects. The Right to Food Unit (Right to Food Team) was directly involved in some of these country level activities, providing technical support, policy advice, capacity development and advocacy, and a space for multi-stakeholder dialogue.

Through effective action in the five areas of activity dealt with by the panels, countries will give voice to the hungry, ensure accessible justice, enhance the effectiveness of its institutions and achieve durable impact. The outcome of these interventions should constitute a significant contribution to sustainable development and the achievement of the MDGs.

Topic 1. Strong Voices: Advocacy and Training

Issues

The complete or partial failure by duty bearers to carry out their tasks and responsibilities is not necessarily due to lack of political will but often to lack of capacity, knowledge, skills or experience to undertake the required tasks. With increased knowledge and better understanding, inter-personal communications should improve, together with the capacity to make decisions. When individuals are well informed about the root causes of hunger and appropriate ways of addressing these causes, they may be more motivated to take action and feel more secure in accepting responsibilities.

Gabriele Zanolli / FAO

It is essential to provide public officials with the necessary training and awareness so that they have a full understanding of what the right to food means in practice and are thus in a position to fulfil their obligations with respect to this right, both effectively and efficiently. Right holders require knowledge and understanding in order to hold public officials accountable for the outcomes of their decisions, the management of public resources, or lack of respect for the rule of law.

Empowerment of right holders will increase their capacity to demand their rights and effectively participate in food and nutrition-related decisions that concern them.

Education and awareness raising are addressed in Right to Food Guideline 11. The importance of capacity building among government and judiciary professionals, civil society organizations, universities, primary and secondary schools, the business community and the media, is recognized as a crucial prerequisite for the realization of the right to adequate food. The Right to Food Guidelines offer practical suggestions for states to invest in human resources in order to ensure good health, sustainable resource development and increased educational opportunities at all levels for their people, with a particular focus on girls, women and underserved population groups.

Human rights education and training are key issues to take into account when creating a political, social and institutional environment conducive to the realization of the right to food. Education and awareness-raising programmes and campaigns should target professionals in these spheres, as well as the general public, and especially those affected by food insecurity and malnutrition. Public awareness is an important prerequisite for successful public campaigns.

Capacity building is also required for international agency staff: they need to have full knowledge and understanding of what the right to food means in practical terms, and how it relates to their responsibilities in supporting national efforts for its implementation.

Challenges

The key challenges relate to the nature of information and communications work carried out: this requires thorough technical knowledge of the human right to adequate food and of practical ways for its implementation. Formal and non-formal education require considerable support and time for revision of curricula, the training of instructors, administrative personnel and trainers in general; time is also needed for the production of relevant educational materials (teachers' guidelines, training and learning materials). This is especially true when providing training in economic, social and cultural rights (ESCR), a topic which, until recently, was largely confined to the legal profession.

Now that economic, social and cultural rights have entered the political, economic and social arena, information and communication strategies and materials need to be targeted to different audiences. Messages should relate to the primary concerns of these different audiences and, most importantly, should lead to follow-up action. Consequently, the translation and adaptation of advocacy and training materials into national and local languages is essential.

Formal education and professional training on the right to food needs to be supported by concrete action at country level. Education and training are long-term processes, therefore, all such efforts should have a long-term perspective, be multidisciplinary due to the diverse nature of the tasks to be undertaken, and involve multiple actors.

Country Experiences

Advocacy and training campaigns are taking place in Brazil, Mozambique, India, Philippines, Guatemala and other countries, to provide both government officials and the general public with a better understanding of the right to food and its implications.

Advocacy:

Philippines: In the Philippines, people refer to the right to food as 'the justice of eating'. ESCR-Net Asia, which is an NGO advocating for ESCR, is working with FAO to include human rights in the Philippines' hunger mitigation programme. By expanding the knowledge base of both government and civil society on the right to food, the ESCR-Net Asia team is managing to increase the number of stakeholders involved. Toolkits containing a 'how-to' manual along with educational CDs and video programmes have also been developed, for use by policy makers.

Guatemala: The country has all the necessary laws, policies and strategies at hand to implement the right to food. However, implementation is critically dependent on training and effective advocacy. One project in particular, supported financially by FAO, has conducted several training courses for officials involved in government, civil society and the UN. The right to food has been incorporated in the curricula of two universities in Guatemala: San Carlos University and Rafael Landivar. Young people have been engaging in a 'community coexistence' programme where groups of students live for three days and two nights in households that are vulnerable to food insecurity. This endeavour has been very successful; the students greatly increased their knowledge and awareness of the issue. They then contacted the government officials responsible for work at community level and worked together with them to instruct on and familiarize them with developments and current legislation on the right to food and its implications.

This particular awareness raising initiative has resulted in the setting up of several projects aimed at community development.

Global campaigns to promote the right to food include: 'Hunger Free' by ActionAid, 'Face-It-Act-Now' by FIAN, the *Campaña sobre el Derecho a la Alimentación* by PROSALUS, and 'The Right to Food: Make it Happen – World Food Day 2007' by FAO. Civil society initiatives have also been set up, such as the 'African Network for the Promotion of the Right to Food'. In order to reach people from all walks of life in promoting the right to food, PROSALUS in Spain has utilized theatres, puppet shows, storytelling, posters, DVDs, CD-ROMs, websites, conferences and university discussion groups, as well as disseminating publications on related issues, such as biodiversity, biofuels and trade.

Education and Training:

Uganda: For the past ten years now, Makerere University in Uganda has been offering a right to food course as part of its Master of Arts programme in human rights. The programme, which is aimed at teachers of law and the social sciences, was developed in the aftermath of war, at a time when many people in government and civil society were talking about human rights but few fully understood what such rights meant in practice. M.A. graduates from this programme are now working in civil society, in government or in the national human rights commission, among other institutions. The University is about to publish a second volume of the African Journal on Ethics and Human Rights, entitled 'The Right to Food and Development in Africa'. This is part of an effort to strengthen capacity in right to food research, and develop a right to food curriculum with learning modules on specific themes.

Brazil: The civil society organization Brazilian Action for Nutrition and Human Rights (ABRANDH), together with the Ministry of Social Development and the Fight Against Hunger (MDS), has developed a distance learning course on food and nutrition security. The course was set up to train social workers directly or indirectly involved in human rights, food security and nutrition, so that they can effectively promote the right to food. When enrolment commenced for the first course in 2007, 5 232 people (comprising government staff, personnel from civil society organizations and university students) signed up and approximately 50 percent completed the course.[19] The general opinion was that the course was both useful and practical and most of the participants said that they would recommend it to others.[20] ABRANDH has also been working on capacity building with leaders in two vulnerable communities in Northeastern Brazil. Its aim is to undertake an assessment of community problems and help the communities to understand that these issues constituted a violation of their rights, including their right to food, all of which needed to be addressed. Community leaders were trained to dialogue and negotiate with local government on how to solve the problems they had prioritized. This endeavour empowered the people and developed their leadership skills.

Europe: An increasing number of universities in Europe are currently engaging in right to food education. These initiatives frequently take the form of introducing right to food modules in

---

19 http://www.direitohumanoalimentacao.org.br/portal

20 A second online course was held in 2009, completed by 1700 persons throughout the country, and further on-line trainings are planned in collaboration with the MDS. The contents of the courses are reflected in a manual developed by ABRANDH (Available at http://www.abrandh.org.br).

Master's Programmes in human rights. Examples of this are the National University of **Ireland**, Galway, and La Sapienza University in Rome, **Italy**. Both the University of Oslo and Akershus University College in **Norway** are offering full courses on the right to food.

Conclusion

The panel discussions indicate that building awareness and understanding of the right to food through advocacy and training is a necessary step prior to and during the development of policies promoting the right to food. The right to food empowers stakeholders – be it right holders or duty bearers – and spurs them into taking responsible action. Thus right holders become capable of demanding their rights or creating the environment in which those rights can be exercised. This approach increases the effectiveness of agencies working at country level that may not otherwise have capacity or expertise to promote human rights. The challenges are to create information and educational materials relevant to each category of stakeholders and accessible to all in a timely manner through formal and non formal channels of knowledge dissemination. These include global campaigns of advocacy, formal university level coursework, training courses for public officials and citizenry and distance learning tools for all.

Strong Voices: Advocacy and Training

Lessons Learned:

1. Training and advocacy are paramount to ensuring the inclusion of all stakeholders in the right to food policy making and implementation process.

2. Universities play a vital role in education and dissemination efforts to reach experts and staff in government institutions and civil society organizations. They also build academic interest in human rights and in the right to food.

3. New information and communication technologies, including e-learning methods, virtual classrooms and free media outlets, have become popular and valuable tools for sharing right to food information. E-learning techniques are very effective in reaching large numbers of people.

Recommendations:

1. It is recommended that education and advocacy material be targeted and customized based on identified stakeholders and their needs. Training and advocacy must not be limited to one segment of a population. Public officials, civil society and development agency personnel should be equally considered in efforts to educate people on the right to food.

2. The impact of formal academic courses on knowledge dissemination and human rights capacity development should not be overlooked. All efforts by universities to include the right to food in course curricula should be strongly supported.

3. New information channels and specific e-learning tools should be maximized to increase the impact of right to food information, knowledge and messages.

Topic 2. Accessible Justice: Legislation and Accountability

Issues

It is only as a result of enforceable justice, trusted institutions and a legal system oriented towards the human right to food that right holders will be in a position to hold duty bearers accountable for guaranteeing food security.

Accountability is intended to mean the obligation of those in authority to take responsibility for their actions. In the right to food context, four main types of accountability mechanisms are of particular relevance: political, administrative, legal and social. The main concern here is accountability through the application of law by government, independent human rights institutions and courts of law.

Gabriele Zanolli / FAO

Right to Food Guideline 7 addresses the legal framework and envisages possible domestic legal and constitutional provisions on the right to food. It stresses the need for adequate, prompt and effective remedies in situations where rights are not upheld, and states that right holders need to be informed about their rights and methods of claiming them.

With regard to legislation on the right to food, it is important to distinguish between four related issues:

1. *Constitutional recognition and protection of the right to food*, which can be explicit or implicit, listed in the bill of rights or as a directive principle of state. The protection of the right to food in the constitution is the strongest form of legal protection, as constitutions are considered to be the fundamental or supreme law of a country.

2. *Framework law*, which spells out rights and obligations in greater detail than is usual in constitutional provisions, and is important because it indicates an institutional framework for implementation, monitoring and further action. It also helps officers of the courts to tackle right to food violations.

3. *Sector laws*, may either help or hinder the implementation of the right to food; consequently, they should be reviewed for compatibility with the right to food and with human rights principles. Sectoral legislation is important because it regulates the economic environment in which people are or are not able to feed themselves in dignity; for example, it regulates the adequacy of the food that is marketed and sold and regulates access to the natural resources by which people feed themselves.

4. *Justiciability* of the right to food makes it possible for an individual to lodge a complaint before a court or other independent authority regarding a violation of his or her right, and to obtain appropriate means of recourse and remedy. This is important because laws that are

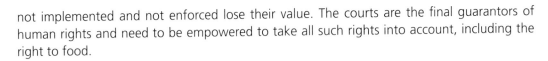

not implemented and not enforced lose their value. The courts are the final guarantors of human rights and need to be empowered to take all such rights into account, including the right to food.

Challenges

Legislative processes are complex undertakings, involving different and often diverging interests, and leading to outcomes that reflect what was achievable under prevailing circumstances, rather than identifying best possible solutions. Governments and legislators are often wary of adopting 'legislation with teeth'; they are more comfortable with laws that do not imply major constraints or obligations. The lack of political will may be due to fear of legal accountability: in fact, only a few countries argued for strong guidelines on legislative measures during negotiations on the Right to Food Guidelines.

Legislation, however meaningful, may not always be implemented in practice. This is a common problem, especially in poorer countries, and one that easily leads to a lack of respect for the rule of law. Apart from the problem of non-implementation, judicial systems in many countries suffer from backlogs, high court costs, corruption and other challenges.

Experience with framework law is very recent and still rather limited. Evidence is not yet available as to whether or not these laws have struck a successful balance between making a difference and being realistic.

Rights have to be balanced against other rights. Frequently, this tug of war cannot be addressed by law but by policy decisions, reflecting power relations that are often biased against the poor. In such a context, poor people's rights, such as the right to food, are likely to be considered less important than the rights of more influential groups, such as property rights or the right to set up a business.

The enjoyment of the right to food can be affected positively or negatively by increased investment in areas such as mining, biofuels and large-scale, intensive agricultural production. In such cases, new jobs are created but self-employment and poor people's access to land and other natural resources may be threatened in the process. These developments are facilitated by investment codes and mining laws, which are usually 'legislation with teeth', and should therefore be reviewed as a priority, to ensure compatibility with the right to food. The challenge is to create clear and transparent rules that make provisions for public consultation, compensation for land loss and benefits to the community where the investment is made. A range of tools are currently being developed in several countries to ensure that changes are for the benefit of those most in need.

States' international obligations may be in conflict with one another, such as the right to food versus intellectual property rights or other trade-related obligations. In the current system, taking into account both the interests at stake and the enforcement systems, trade interests are more likely to prevail.

Lawyers and judges have little or no experience in using the right to food as an argument in legal cases. This capacity gap needs to be addressed through law schools and on-going training and updating of judges and lawyers on right to food matters. The scarcity of precedence and jurisprudence is also a constraint for cases to be argued and won on the basis of the right to food.

Poor people have very little access to justice, in practice. They are unaware of their rights and have no knowledge of the law. Due to the high cost of legal representation, they cannot afford to approach lawyers who are able and willing to take on their cases; sometimes they may even consider it disrespectful to challenge authority.

Country Experiences

Constitutions:

Most national constitutions throughout the world recognize the right to food or at least some aspects of it. Many have general provisions on the right to an adequate standard of living, the right to a dignified life or the right to development, all of which include the right to food implicitly, if not explicitly. A number of constitutions contain provisions relating to the right to food as non-justiciable directive principles of state. The right to food as an explicit human right that is justiciable in court is recognized in 20 national constitutions only, of which 10 countries specifically recognize children's right to food.[21]

South Africa has clear constitutional provisions regarding the obligations of all organs of the state to respect, promote, protect and fulfil the right to access food. By contrast, **India** cites nutrition as a directive principle of state only, while the right to food is protected by a broad interpretation of the right to life.

Bolivia and **Ecuador** have given prominence to the right to food in recent constitutional reforms. The constitutional court in Bolivia applies international agreements and the new constitution is a strong legal tool for the protection of the right to food[22]. In Ecuador, individuals' right to food is fully recognized and clearly justiciable in its new Constitution[23]. In addition, there are provisions regarding equal access to means of production and non-discrimination.

The 1995 Constitution of **Uganda** acknowledges, under its section entitled *National Objectives and Directive Principles of State Policy*, that all Ugandans [should] enjoy access to food security, and that the State shall take appropriate steps to encourage people to grow and store adequate food.

Framework Laws:

Since the adoption of the Right to Food Guidelines in 2004, many countries have started to draft framework laws, primarily in Latin America, and also in Africa. **Brazil** and **Guatemala** have adopted framework laws that bring coherence and stability to their food security system and recognize the right to food as a human right. At the time of the Right to Food Forum, **Ecuador**, **Honduras**, **Mozambique**, **Nicaragua**, **Peru** and **Uganda**, among others, were in the process of drafting new framework laws on food security and the right to food.[24]

---

21   FAO. 2009. *Guide on Legislating for the Right to Food*. Rome. See also, FAO's forthcoming publication *Constitutional and Legal Protection of the Right to Food around the World*. Rome. 2010. The latter shows that by the end of 2009, 22 national constitutions explicitly mention the right to food as a human right. (All publications as well as a legal database related to the right to food available at http://www.fao.org/righttofood).

22   http://www.fao.org/righttofood/kc/legal_db_en.asp?lang=EN

23   See Chapter 2, Section 1, Article 13 of Constitution of Ecuador.
     (Available at http://www.asambleanacional.gov.ec/documentos/constitucion_de_bolsillo.pdf).

24   The National Assembly of Ecuador approved a law on food sovereignty in February 2009 entitled *Ley Organica de Regimen de Soberania Alimentaria* (Available at http://www.asambleanacional.gov.ec/documentos/leyes_aprobadas/ley_soberania_alimentaria.pdf).

Brazil took an important step in ensuring the human right to adequate food by introducing the Federal Law for Food and Nutrition Security (LOSAN, Law No. 11346) in 2006. This framework law on the food security system stipulates that adequate food is a basic human right, inherent to human dignity and indispensable for the realization of the rights established by the Federal Constitution. The law states that the government shall respect, protect, promote, provide, inform, monitor, supervise and evaluate the realization of the human right to adequate food, as well as guarantee the implementation of specific claim and recourse mechanisms. It creates an institutional framework, including a large advisory body, the National Council on Food and Nutrition Security (CONSEA) and an inter-ministerial coordination mechanism. The national food and nutrition security system seeks to formulate and implement policies and plans on food and nutrition security, motivate the integration of government and civil society endeavours, and promote the assessment, monitoring and evaluation of food security and nutrition throughout the country.[25]

Guatemala implemented the recommendations of General Comment 12 of the Committee on Economic, Social and Cultural Rights (CESCR) when it promulgated Legislative Decree 32-2005 establishing the National Food and Nutrition Security System (SINASAN). The enactment of this law and the adoption of a State Food Policy in 2005 constituted a major step forward, opening up greater possibilities for government to take on the responsibility to respect, protect and fulfill the right to food.[26] Subsequent experience has demonstrated that the adoption of a law does not automatically lead to change: what is needed is people's commitment, the means to implement the laws and a strategy by which to do so.

In **Nicaragua**, work on a food security and sovereignty framework law started in 1996. A draft bill was ready in 2001 and in June 2009 the National Assembly passed the *Ley de Sobernia Y Seguridad Alimentaria Y Nutricional* which explicitly recognizes the right to food[27]. It takes into account the multidimensional nature of food and nutrition security, and provides for the participation of different public and private institutions as well as civil society. The law establishes a National System for Food and Nutrition Security and Sovereignty (SINASSAN) to implement the right to food. The system includes a National Commision for Food and Nutrition Security and Sovereignty (CONASSAN) to coordinate intersectoral and inter-ministerial efforts at the national level. The law also provides for regional and municipal institutional mechanisms to implement it.

Venezuela has undertaken a complete reworking of its legal framework for the right to food. The process began with the 1999 Constitution in which food sovereignty and food security were enshrined. In July 2008, a framework law on food security and sovereignty[28] was passed by the National Assembly. It regulates proper access to food with strategic reserves and planning of these reserves, fair trade and exchange, safety and quality of food, and also education and training in nutrition principles.

---

25 See present publication – Part THREE, Country Case Studies: I. Brazil – A Pioneer of the Right to Food.
    (The law, including the English translation and legislation material are available at http://www.fao.org/righttofood/
    kc/legal_db_en.asp?lang=EN).

26 See present publication – Part THREE, Country Case Studies: II. Guatemala – Writing a Piece of History.

27 Ley No. 693. *Ley de Soberenia Y Seguridad Alimentaria Y Nutricional 2009* (Available at http://www.asamblea.gob.ni).

28 Decreto 6.071. *Ley Organica de Seguridad y Soberania Agro-Alimentaria 2008* (Available at http://www.asamblea.gob.ni).

Uganda's draft Food and Nutrition Bill, which is being prepared on the basis of explicit provisions in the country's food and nutrition policy, states that everyone has the right to food and to be free from hunger and undernutrition. It defines the right to food in accordance with General Comment 12 of the CESCR and has strong non-discrimination provisions. It puts emphasis on vulnerability, on the grounds of age, health, displacement, etc. and states that the government should provide a minimum food entitlement to persons who cannot feed themselves. The draft provides a legal basis for a Food and Nutrition Council, which is tasked with implementing the policy and the law. There are also provisions on government accountability and on recourse mechanisms.

In **Mozambique**, the Ministry of Agriculture, in collaboration with the Ministry of Justice, hopes to have a right to food bill ready shortly, as foreseen in the country's poverty reduction plan.[29]

Sectoral Laws:

In **Bolivia**, the government is opposed to competition from strong business interests and has taken steps to protect people's right to food by restricting commodity exports. It is strengthening national companies so that they can provide supplementary foodstuffs without having to look abroad for extra supplies. It has also inserted the Right to Food Guidelines into the country's development plan and created a National Council for Food and Nutrition Security (CONAN) comprising a wide selection of people from all government ministries.

Brazil at the time of the Forum was working on a bill that governs the provision of school meals.[30] That bill was to cover both public and private schools, and extend the school lunch entitlements to older children, beyond elementary school. It was also to provide for at least 30 percent of food purchases to be sourced from family farms.

The National Rural Employment Guarantee Act of 2005 is one example of an important food security instrument in **India**. Part of the country's policy on food security also includes the public expenditures on food safety nets such as the Public Distribution System (PDS), the Mid-Day Meal scheme and the Integrated Child Development Services (ICDS). ICDS entitles adolescent girls, pregnant and lactating women and children under six to take-home rations or cooked meals. However, legal provisions in India have to comprise more specific and detailed entitlements for children in different age groups. There is also a need for stronger provisions with regard to maternity benefits, crèches, breast milk substitutes and the protection and promotion of breastfeeding. In addition, India is planning a National Food Security Act, expressly aimed at protecting the right to food.[31]

---

29 A task force is working on the legislation that includes government, civil society, academics and other stakeholders. There are regular meetings in Maputo each month, to go over the process, discuss the contents and elicit as many ideas as possible. With widespread participation from the beginning to the end of the legislating process, all concerned are likely to be familiar with the proposed legislation; consequently the final law can be implemented more easily.

30 For further information refer to current *Law no. 11.947/2009*.

31 In June 2009, representatives from different civil society organizations met to discuss the first draft of this Act, and to formulate 'essential demands' for the Right to Food Act. See the Right to Food Campaign, Right to Food Act, Introduction (Available at http://www.righttofoodindia.org/right_to_food_act_intro.html).

Justiciability:

Jurisprudence on the right to food is still limited, therefore, justiciability of the right to food has not been tested in courts of in most countries. This does not mean that the right to food is not justiciable *per se* but that it has rarely been argued successfully in court. However, in cases where the right has been argued for and defended, existing jurisprudence shows that some judges have made it obligatory to guarantee the right to food with no form of discrimination.

Uganda's judicial system comprises a hierarchy of judicial and administrative redress mechanisms starting with Local Council Courts, followed by the Family and Children Courts, the Tribunals, the High Court, and ending with the Supreme Court. In the case of infringements, the recently elaborated Food and Nutrition Bill (2008) provides for the following: penalties (imprisonment, fines), restitution, cessation of the unlawful act, guarantee of non-repetition, or compensation. It is necessary to strengthen rural poor people's access to remedies and representation. The Uganda Law Society has launched a legal aid project concerning civil and political rights.

In **India**, the People's Union for Civil Liberties filed a public interest petition in the Supreme Court in 2001. This case became famous in India as the 'Right to Food Case'. The Supreme Court had paved the way for the implementation of the right to food when it explicitly stated in various judgments that the right to life should be interpreted as the right to live in human dignity, which includes the right to food. Furthermore, it found in interim orders that the prevention of hunger and starvation *'is one of the prime responsibilities of the Government – whether central or state.'*[32] This case turned out to be the largest class action petition ever filed in terms of its impact upon people. More than 20 interim orders have been issued since 2001. Although described as 'interim orders' these are, in fact, final orders in the specific area they cover. The Indian court system is unique in that it is so broadly defined as to allow almost anyone to make a petition on behalf of people too poor or illiterate to do so themselves. Written authorization is rarely insisted upon. The public interest petitioner is presumed not to have sufficient resources to collect the data needed for the conduct and successful prosecution of the case. In 2002, the Supreme Court appointed two special commissioners to collect information and monitor compliance with its orders.

The success of the case in India lies in having a movement to back up the formal proceedings and maintain public pressure. The painstaking costing of entitlements and collection of data enabled the Supreme Court to be precise in its orders.

In **South Africa**, the right to food is included in the Constitution of 1996. A case was heard in the High Court in the Cape of Good Hope, where thousands of fishermen had lost access to the sea because of a 1998 law on fisheries (Marine Living Resources Act). The Law established quotas, which meant that only commercial fishermen could fish. This resulted in the minor fishermen having no access to the sea for six years. In 2004, the fishermen made an appeal through an organization for development in the Cape of Good Hope, arguing that implementation of the law violates their right to food. The UN Special Rapporteur on the Right to Food filed an affidavit before the courts, explaining what is meant by the right to food and how the right should be used in this specific case. Upon examining the case, the Court ratified an amicable agreement between the Government and the fishermen allowing 1 000 fishermen to have immediate access

---

32   Supreme Court Order of 20 August 2001, *PUCL v Union of India and others*, Writ petition (civil) 196 of 2001.

to the sea. In a judgement given in 2007, the Court took the responsibility for ensuring the implementation of the friendly agreement and obliged the government to review the law and respect the right to food in this case.

Pro public, a non-governmental organization, was spurred into action during the recent food crisis in **Nepal** by news reports of people starving and stealing food. They brought a case before the Supreme Court on the basis that the Nepalese Constitution provides for the right to food. Referring to WFP's vulnerability assessments, which revealed that 42 out of 75 districts of Nepal were food insecure, the Court ordered the government to provide for food supplies to 12 of the 42 districts.

In 1995, two undocumented refugees in **Switzerland** went to court after being refused social assistance in accordance with existing legislation. The Supreme Court acknowledged that anyone living in a democratic state had the right to a minimum level of subsistence, based on human dignity. This was an unwritten constitutional rule. Since then, the Swiss Constitution has been amended, explicitly recognizing the right of everyone to a minimum level of subsistence.

In **Brazil**, like in many other countries, the process of claiming the right to food is both complex and difficult; the courts are often inaccessible for the poor. Without civil society involvement in the monitoring and reporting of rights violations, duty bearers may remain ignorant of cases of relevance to right to food.

One important instrument is cooperation with the Public Prosecutor who is totally independent of the executive, legislative and judicial branches and represents society before these three branches. The Public Prosecutor can convene a public hearing to investigate individual or group claims of rights violations and can facilitate the adoption of remedial measures. When a case involves conflict between the state and society, the Public Prosecutor has the authority to summon government bodies and other relevant parties to a hearing and negotiate a Terms of Conduct Adjustment. This is a legally binding settlement that defines obligatory actions for all parties and establishes a time-frame within which the identified violations have to be corrected. Public civil suit is considered a last resort.

Conclusion

There are a variety of legislative and policy instruments through which a country can advance the right to food once the commitment exists as a national priority. There are issues to consider even after a law or policy protecting the right to food is in place, such as whether:

- the right is enforceable;
- remedies provided are adequate and effective;
- the law itself is applied justly;
- and officials in political, administrative, legal and social spheres perform their duty to make the right accessible.

In particular, courts must be empowered enough to guarantee the right to food just as much as people must be informed enough to be able to make a claim. Finally, in the case of overarching policy frameworks covering the right to food, there must be coherence with policies related to the right to food and relevant to all sectors of the economy. This is to say that sectoral laws and their application must be aligned with the right to food objective.

Accessible Justice: Legislation and Accountability

Lessons Learned:

1. Although the strongest form of protection of the right to food is a constitutional recognition of the right, framework laws may offer a higher level of protection by outlining the implementation, monitoring and claim mechanisms for the right to food. Furthermore, in some cases sectoral laws can hinder the implementation of the right to food.

2. Countries that show strong political will and commitment to the right to food are the ones most successful in implementing policies and laws aiming at promoting this right.

3. In implementing the right to food the actual process is just as important as the textual provisions of a law. There is a clear need to strengthen accountability mechanisms and local access to justice even where the right to food is recognized. Information regarding existing channels to file a complaint or ways to claim a justiciable right to food must be imparted and accessible in a meaningful way.

4. Lawyers and judges still have little experience in right to food although they have the key role of representing right holders and creating functioning judicial systems that offer true remedies in case of violations.

Recommendations:

1. Countries should review and strengthen their legislative framework on the right to food through all three main instruments: constitutional protection, framework and sectoral laws. Compatibility review of all sectoral laws with the right to food objective is necessary as recommended by the FAO Guide on Legislating for the Right to Food.

2. Processes that support political expression of interest and real commitment to right to food should be supported.

3. Informing citizens on rights and how to claim them should be a priority followed by the commitment to make available administrative and judicial recourse mechanisms. Investigations on judicial systems must be considered to identify cost restrictions, corruption and unnecessary delays in processing claims that discourage right holders from claiming their right to food.

4. Judges and lawyers in particular must be targeted through awareness raising campaigns promoting the right to food. Their capacity to handle right to food cases must be addressed in both law schools and ongoing professional legal trainings. It will be necessary to train NGOs and CSOs at local level to create the possibility for developing public interest litigation and creating other means of facilitating access to justice.

Topic 3. Right Targets: Information and Assessment

Issues

By ratifying or acceding to the International Covenant on Economic, Social and Cultural Rights (ICESCR), 160 countries[33] have accepted the obligation to respect, protect and fulfil the right to food. They should therefore create and maintain an enabling and coherent policy, legislative and regulatory environment that enables individuals to produce or procure sufficient, safe and nutritious food for themselves and their families in order to conduct an active and healthy life. States should also provide direct support to those who are unable to feed themselves. Action to comply with these obligations can take various forms: drafting policies and strategies, implementing programmes, adopting laws and regulations, as well as establishing and strengthening institutions. Existing policies should be assessed to see whether policy objectives, goals and strategies are consistent with furthering the right to food, and whether they are implemented in ways that respect human rights principles.

Gabriele Zanolli / FAO

For any right to food action to be taken at country level, one basic requirement is adequate information and assessment of the right to food situation. If the government does not know which people are food insecure and vulnerable, and why they are deprived of their right to food, no remedial action can be determined and effectively implemented to deal with the situation.

The Right to Food Guidelines contain a number of recommendations that help to identify the food insecure and vulnerable. The Guidelines 13 and 14 stress disaggregation of data and identification of vulnerable groups and individuals with a view to taking corrective measures and adequately targeting assistance in relation to needs. Participatory assessment of the economic and social situation and food security is recommended in Guideline 2.2 and institutional assessment in Guideline 5.1.

In referring to strategies for the implementation of the right to food, Guideline 3.2 recommends that duty bearers start with '...a careful assessment of existing national legislation, policy and administrative measures, current programmes, systematic identification of existing constraints and availability of existing resources'.

Five important areas should be considered when undertaking a right to food assessment:

1. Identifying the food insecure and vulnerable.

---

33   Status as of December 2009.
        (Available at http://treaties.un.org/Pages/ViewDetails.aspx?src=TREATY&mtdsg_no=IV-3&chapter=4&lang=en).

2. Assessing the legislative framework to find out whether the legal environment is conducive to realizing the right to food.

3. Assessing the policy framework to analyse whether the formulation and implementation of policies and programmes are consistent with human rights principles, and whether national policies address the underlying causes of non-realization of the right to food.

4. Assessing the institutional frameworks and civil society participation to obtain information about the mandates and performance of relevant public institutions, the coordination mechanisms and the ways to claim the right to food.

5. Assessing budgetary allocations and expenditures, and applying budget analysis techniques, to gauge levels of political commitment to realising the right of food.

Challenges

Lack of disaggregated data, or inaccessibility of such data in many countries, limits the possibility of identifying food insecure and vulnerable groups according to livelihood or socio-economic characteristics. This constitutes an obstacle to better understanding the underlying reasons for their being food insecure or vulnerable.

A major challenge is to identify the relevant legal issues, policies, institutional arrangements, and administrative measures to be included in a right to food assessment. When time is short and assessment resources limited, there will need to be some prioritization in order to focus on the most important and relevant aspects (such as the need to define the assessment sphere).

The work of FAO, Action Aid, Rights and Democracy, FIAN and others have generated some useful insights as to the merits and difficulties of conducting a right to food assessment at country level. The available tools take the assessment beyond food security analysis by linking the information to a country's policies, laws and institutions. Given the complexity of the issue, it is still a challenge to identify the most important programmes or laws to be examined and to analyse them in human rights terms. Most assessment reports tend to be too descriptive, with too much detail and too little analytical depth. For example, they might state that right to food is not included in a country's constitution but fail to analyse how this shortcoming hinders people's ability to realize their right to food or, alternatively, how a constitutional amendment could facilitate the realization of that right.

In-country capacity to undertake a right to food assessment may be limited in many countries, initially, and therefore capacity strengthening is paramount. Since, ideally, the right to food assessment should involve both right holders and duty bearers, the capacity strengthening tasks may become a large undertaking, requiring time and resources.

International development partners may see human rights work as being impractical, or may question the relevance of such rights. Reluctance on the part of donors to apply the human rights framework in hunger eradication policies, for example, often dampens government interest in engaging in the assessment process.

Complementary to a technical right to food assessment, as outlined above, is an analysis of government openness to the right to food and whether there is a human rights culture – meaning that human rights are acknowledged and respected and that duty bearers are mindful of their human rights obligations. This has been demonstrated by FIAN's documentation and analysis of concrete right to food violations occurring in several countries.[34]

Country Experiences

Identifying the Food-Insecure and Vulnerable:

In India, more people die from malnutrition each year than those who lost their lives (some 3 million people) during the 1943 Bengal crisis – India's last major famine.[35] Far too many of them are children. Government surveys typically classify the population into income quintiles, which sheds light on poverty and inequality but fails to identify groups such as the Adivasis and Dalits who face social discrimination with an impact on food security. Disaggregated data is essential in order to handle this problem. The commissioners of the Supreme Court, while deploring the lack of official data, have found plenty of anecdotal evidence of discrimination and social exclusion. They have urged that data collection methodologies be revised to address the information gap and better identify the most vulnerable sections of society.

In **Uganda**, disaggregated data on the number and distribution of vulnerable groups (i.e. children, women, the elderly, the chronically sick, ethnic minority groups, persons with disabilities, and others lacking adequate means of access to food) is lacking, in spite of the fact that vulnerable groups have been identified in policy documents. The Ministry of Agriculture, Animal Industry and Fishery (MAAIF) and the Ministry of Health (MOH) undertook a joint assessment of the food and nutrition situation in the country in 2004 to provide background information for the formulation of the Uganda Food and Nutrition Security Strategy and Investment Plan (UFNSSIP)[36]. Poverty-related nutrition vulnerability – the likelihood of a person succumbing to risk – was reported in the context of the Poverty Status Analysis undertaken by the Ministry of Finance, Planning, and Economic Development (MFPED) in 2003 and again in 2005. Vulnerability has been categorized in both assessments, based on the risk of exposure to armed conflict, as well as on demographic and poverty-related factors.[37]

Information gathered by health teams during vaccination campaigns and by the Indigenous Health Group of the Brazilian Health Foundation yielded the most comprehensive population profile **Brazil** has ever produced. Disaggregated data made it possible to draw up a profile by region, ethnicity, race, gender and age, thus enabling the identification and mapping of specific populations and regions most vulnerable to food and nutrition insecurity. This is crucial, for example, when needing to address the particular dietary needs of specific groups. Brazil's Unified Register facilitates targeted hunger reduction strategies because the MDS now knows exactly which people are poor, why they are food insecure, what they need, and where they live.

In **Guatemala**, the Food and Nutrition Security Information and Coordination Centre (Centro de Información y Coordinación de Seguridad Alimentaria y Nutricional – CICSAN) created the

---

34   www.fian.org/cases

35   WFP. 2006. *World Hunger Series Report – Hunger and Learning*. Rome.

36   Republic of Uganda. 2005. *The National Food and Nutrition Strategy, Final Draft*.
     (Available at http://www.health.go.ug).

37   This category was initially reported in the 2003 Poverty Status Report by the MFPED. It was also adopted in the Uganda
     Food and Nutrition Strategy and Investment Plan (UFNSIP) of 2005.

Municipal Information System on the Risk of Food and Nutrition Insecurity. This information system is designed to collect information on several vulnerability factors, making it possible to rank municipalities according to levels of vulnerability, so as to help the geographical targeting of governmental action.

Assessment of Policies, Laws and Institutions:

In **Brazil**, CONSEA's Standing Commission on the Right to Food recently developed an assessment tool for generating information to advise the government on how to incorporate the right to food into public policies. The policies are to be examined to see if they meet the following criteria:

* Right holders and duty bearers are clearly defined and can be identified,
* Right holders are empowered, and are encouraged to participate in public debates,
* The state respects, protects and fulfils the right to food,
* Mechanisms to claim the right to food are created and have been strengthened,
* Goals, and time-bound targets and benchmarks are defined and monitored,
* Strategies are in place to disseminate information,
* Capacity is being strengthened for both duty bearers and right holders.[38]

Mozambique's Poverty Reduction Strategy 2006-2009, known as Plano de Ação para Redução da Pobreza Absoluta 2006-2009 (PARPA II), adopts a right to food approach, to be implemented through, among others, the formulation and approval of a Right to Food Law. In response, the Technical Secretariat for Food Security and Nutrition (SETSAN) assessed the country's relevant legal framework. Preliminary findings indicate that the right to food is not clearly recognized in national legislation, nor is a human rights based approach applied in the related laws that have been enacted.

The National Anti-Poverty Commission of the **Philippines** organized a multi-sectoral group to meet and discuss the root causes of hunger, the legal framework for hunger reduction, and current social safety net programmes. Three assessment reports were prepared based on nation-wide consultations with NGOs, people's organizations and the private sector. These reports were reviewed by members of national and local governments, civil society and the private sector. This led the government to formulate a national food policy to better coordinate and integrate interventions because it was found that existing laws had resulted in uncoordinated sector programmes and other constraints.

The Government of **Bolivia** is trying to eradicate undernourishment, which affects one in every three children under five, by targeting specific municipalities known to be vulnerable to food insecurity and health problems.[39] Each municipality is authorized to tailor the programme to its own needs. Rather than having one policy for each community, this ruling allows the policies to be drawn up by those who know best how to help their communities.

Malawi, Nepal and Haiti: Rights and Democracy, a Canadian NGO, conducted fact-finding missions in Malawi in 2006, in Nepal in 2007 and in Haiti in 2008. The objective of these missions was to encourage greater accountability on the part of the state and also to empower

---

38   See present publication – Part THREE, Country Case Studies: I. Brazil – A Pioneer of the Right to Food.

39   See Malnutrition Zero program of Bolivia.
      (Available at http://www.unicef.org/bolivia/PR_080708_-_cooperation_agreement_bolivia_canada.pdf).

right holders. The assessment exercise generated new interest in the human right to food as a useful and dynamic tool. The missions, which included both national and international experts, facilitated dialogue among a range of stakeholders who did not normally interact with each other (such as national human rights commissions, agriculture ministries, grassroots organizations and parliamentarians). The endeavour proved to be a success and the dialogue has continued in all three cases. Malawi and Nepal then created national right to food networks to implement the assessment recommendations. Malawi is now moving towards the adoption of a right to food framework legislation, while awareness workshops on right to food issues have been conducted in eight districts of Nepal. In Haiti, a new campaign is now planned to encourage ratification of the International Covenant on Economic, Social and Cultural Rights.

Conclusion

It is obvious that prior to implementation of any right to food law or policy, government officials must have a sound information base in order to identify those segments of society that are most vulnerable to food insecurity and therefore need to have their entitlements to food protected. This should be done in order to create appropriate remedies for each distinct segment of society according to its needs. This process allows correct targeting of campaigns, of resources and programmes, and of other efforts needed to realize the right to food. Assessments are accurate when they involve the full participation of targeted communities in policy programming. It is equally important to carefully examine the national context and the political, legal, social, and policy environments of a country. Such analysis allows the identification of the level of readiness of a country for the implementation of the right to food, that is, whether it has the available resources and what type of constraints exist that might impact the success of implementation efforts. Data collected from both types of assessments should be analysed to find the corresponding links between the two and ensure that duty bearers can be equipped with the right information to direct policy towards a better implementation of right to food objectives.

Right Targets: Information and Assessment

Lessons Learned:

1. Information is essential for duty bearers to identify right holders most in need.

2. Lack of disaggregated data or inaccessibility of data is a severe constraint to identify those vulnerable groups who are most in need of protection.

3. Participatory data collection, particularly with reference to right holders, serves to better understand the causes of hunger. Duty bearers and right holders fully involved throughout all stages of data gathering, increase the likelihood that conclusions drawn from assessments target the right segments and therefore result in the most appropriate benefits and remedies.

Recommendations:

1. Collection of relevant information and regular assessments of both the population and the policy environment should be supported as strategies for effective design and formulation of policy measures promoting the right to food.

2. Policies and programmes should be focused primarily on the most vulnerable and try to rectify discriminatory processes in the implementation of the right to food or lack thereof. This requires the availability of disaggregated data by gender, age, social or ethnic group, etc.

3. Assessments should be initiated by governments preferably in partnership with civil society organizations. Duty bearers and right holders should be fully involved in such tasks at all times, to increase the likelihood that the conclusions and recommendations will be implemented. Information on findings and conclusions should be provided to all members of target groups in a comprehensible and accessible manner, and should be directly linked to solutions and follow up action.

Topic 4. Durable Impact: Benchmarks and Monitoring

Issues

Food security and nutrition goals, targets and benchmarks are usually defined on technical grounds, and from a basic needs perspective; however, they usually lack human rights dimensions. Food security and nutrition monitoring indicators fail to capture the human rights dimensions of realizing the right to food, which means that existing indicators need to be analysed differently and additional rights-based indicators developed. The information required to construct impact indicators needs to be disaggregated so as to analyse the distributional effects of policy measures. Repeated measurements over time of the Gini-coefficient, for example, of distribution of land access or household incomes can indicate whether specific policy measures do, in fact, improve poor people's equitable access to resources.

Rights-focused monitoring tracks the impact of public policies, programmes and actions as well as the ways in which such impacts are achieved. Conventional monitoring indicators usually focus on impacts and outcomes. Indicators such as child stunting, and underweight indicators alone provide little indication as to what policy responses are required. They need to be incorporated in an integrated analysis that links causes to outcomes.

Although causal and vulnerability analysis is being increasingly undertaken at country level, an analysis of the implementation processes is usually not included, hence, the need to develop appropriate process indicators that capture specific elements during implementation and provide clear guidance for remedial action in the case of violation.

Overall, economic growth and the achievement of socio-economic development goals do not necessarily indicate that the human rights of all have been fully respected, protected and fulfilled. This is also true with regard to the realization of the right to food. It is important to monitor the results of state action and also the effect of the process that led to those results. Such processes should adhere to human rights principles and approaches: they should be transparent, non-discriminatory, participatory and empowering, and fully respect the rule of law and human dignity.

Human rights focused monitoring needs verifiable and time-bound targets and benchmarks to be established so as to measure and report progress in realizing the right to food. Such targets and benchmarks should orient policy and help improve governance because right holders and their representatives can use them to hold governments accountable for lack of adequate progress. National targets are increasingly included in policy and strategy documents. This allows others to monitor whether or not these targets are being reached. Guideline 17 of the Right to Food

Guidelines invites States to monitor their progress and to report periodically to the FAO Committee on World Food Security. States that are parties to international covenants and agreements are obliged to report periodically to the respective Treaty Bodies on progress being made towards the realization of the right to food.

Right to food violations need to be monitored so that measures can be taken to remedy such violations and to prevent them from re-occurring. An increase in the number of reported cases over time can indicate a regression in the realization of the right to food. On the other hand, in certain contexts, it could also mean that monitoring mechanisms have been strengthened and are actually better able to identify cases of violations.

Monitoring information should help empower right holders by making it possible for them to hold duty bearers accountable for poor performance, unlawful conduct or wasteful use of public resources. It should help people understand what their rights are, what these rights mean in practice, and how to fulfil their right to food. However, most of this information does not even reach right holders as only restricted technical groups and decision makers have access to such data.

Challenges

Right to Food Guideline 17.1 suggests that *'States may wish to establish mechanisms to monitor... by building on existing information systems and addressing information gaps'*. Existing in-country information systems are often weak, not so much in information availability as in analytical capacity. This restricts right to food monitoring, which requires technical monitoring capacity, full understanding of what human rights principles mean in practice, and the capacity to analyse monitoring information from a human rights perspective. As it is often difficult to find all these skills in one organization, capacity strengthening is invariably required.

Right to Food Guideline 17.6 indicates that: *'...States should ensure a participatory approach to information gathering, management, analysis, interpretation and dissemination'*. Monitoring reports are usually quite technical and accessible to groups of professionals only. However, human rights based monitoring should aim at making monitoring information accessible to and useable by all right holders. This may require monitoring the right to food at community levels and by grass roots groups. It also means that governmental institutions and non-governmental organizations tasked with monitoring the right to food have a duty to make monitoring information accessible to everyone. The monitoring process should be transparent.

Right to food concepts are often misunderstood by decision makers and/or seen as a threat to their decision making power. This constrains human rights focused monitoring of the right to food. Strategies need to be put in place that are designed to provide decision makers and stakeholders with a thorough understanding of right to food concepts, their meaning in practice, and ways in which to apply them in different professional and technical areas.

Monitoring the right to adequate food is a tool still in its infancy. Little empirical evidence is available to date on how to monitor the right to food, what to monitor, for whom and for what purpose. Methods already available and in use for vulnerability assessments, policy analysis, programme and institutional assessments and public budget analysis, should be adapted and applied in right to food monitoring initiatives.

Community-based monitoring is not a new concept. Truly participatory methods have been developed and implemented for many years and there is a body of available literature that synthesises both methods and experiences. Such methods, which often generate qualitative information, should now be incorporated in the monitoring of the right to food as complementary quantitative data.

Country Experiences

Countries are gradually coming to realize the importance of having a single, specific body with a clear mandate for monitoring the realization of the right to food. In order to be effective; such a mandate needs to be recognized by all governmental institutions, and be in accordance with the Paris Principles.[40] Thus, the respective institution needs to have the necessary capacity to undertake right to food monitoring, including the monitoring of right to food violations.

For example, the first such institution in South Asia was the National Human Rights Commission (NHRC) of **India** established by the Protection of Human Rights Act of 1993. Its mandate and independent status comply with the Paris Principles covering national institutions for the protection and promotion of human rights. The NHRC has investigated complaints about starvation in Orissa, a case that has been ongoing since 1997, and farmer suicides in Andhra Pradesh, Kerala and Karnataka. In each case, it appointed a special rapporteur. A section of its annual report is devoted to food security issues. The NHRC monitors particular cases and requests quarterly performance reports on achievements. This was also the case for the districts of Kalahandi, Balangir and Koraput in Orissa. A Core Group on the Right to Food, established in January 2006, provides advice to the NHRC and has recommended the formulation of a national plan on the right to food.

Recently concluded assessments of national food and nutrition programmes in several **Latin American** countries (**Argentina, Brazil, Mexico** and **Panama**) show that it is possible to include a human rights focus in such assessments. These assessments specifically covered the process of programme design and implementation, and the extent to which they are human rights based. Initial findings indicate that national teams require substantial capacity strengthening with regard to: (a) human rights and right to food concepts; (b) their relevance to the programme assessment; and (c) implications for assessment methods. In some countries, the technical assessment team has strengthened its human rights expertise by partnering with a human rights institution or organization.

FAO offered to assist the **Uganda** Human Rights Commission in developing a right to food assessment tool that can be adapted at the district level, to assess the capacity of authorities in identifying and responding to local right to food issues. This tool will be crucial in developing country-specific process, impact and outcome indicators for incorporation in a sound monitoring process at both district and community level. Moreover, FIAN, together with other civil society organizations, has made a significant contribution to capacity building and right to food monitoring in Uganda, by the development and application of a monitoring tool that assesses the implementation of right to food actions contained in the Right to Food Guidelines.

---

40 The Paris Principles were defined at the first International Workshop on National Institutions for the Promotion and Protection of Human Rights in Paris on 7-9 October 1991, and adopted by UN General Assembly Resolution 48/134 of 1993.

The Special Commission for Monitoring Violations of the Right to Adequate Food in **Brazil** is the country's first commission to specifically address an economic, social or cultural right. This Commission, which is housed in the Special Secretariat of Human Rights, investigates claims of violations, and proposes ways for recourse and reparation. When school feeding programmes were jeopardized by pilfered funds, the Special Commission, along with the Ministry of Education and other government agencies, discussed appropriate actions and solutions. Suspending the transfers of funds would have deprived the children of their lunch entitlement so the group suggested several alternatives, including working with public prosecutors, publishing the identities of guilty parties ('naming and shaming') and enlisting local school councils to take part in the supervision of spending. Another non-judicial remedy was to publish the budgets allocated to all state schools. This approach enables parents, teachers and even students to demand that school officials account for the utilization of funds.

As is the case in many countries, the Human Rights Commission of the **Philippines** suffers from a lack of financial and human resources and from a weak mandate on economic, social and cultural rights. While the Constitution limits the mandate of the Human Rights Commission to investigate civil and political rights, it does not, however, limit its recommendatory, research and monitoring powers. Hence, the Commission recently undertook a project to *'develop a common framework for monitoring government's compliance with its obligations on the right to food.'*[41]

Conclusion

Effective monitoring of right to food implementation processes and outcomes determines whether the efforts put forth can and will have an impact that is lasting and fair to all stakeholders. For a long-lasting impact, what is needed is not only a baseline – from which benchmarks are set for the progress in the realization of right to food – but also truly effective monitoring systems and a process that is transparent, non-discriminatory and participatory. Indicators, as a starting point, must go beyond a reflection of vulnerability status and measure distributional impact of policy also. Right to food policies and strategies require a careful analysis of disaggregated data with linkages showing the causes of hunger as it relates to rights, responsibilities, and processes that govern them. In addition to ensuring the outcome of protected human rights, the right to food requires careful monitoring of all stages of implementation to guarantee that human rights principles are respected throughout the process.

---

41   FAO. 2009. *Right to Food Assessment on the Philippines*. FAO/Asia-Pacific Policy Center, pp. 2-36. Rome. (Available at http://www.fao.org/righttofood/publi10/PHILIPPINES_assessment_vol0.pdf).

Durable Impact: Benchmarks and Monitoring

Lessons Learned:

1. Monitoring the process of the realization of the right to food is not just about declaring a political commitment to right to food policies. It implies measuring progress in the implementation mechanisms.

2. Human rights-based monitoring tracks the effect of policies, programmes and actions as well as the process by which these effects come about. It involves producing an integrated analysis of causes and outcomes disaggregated by different levels of food insecurity and by vulnerable segments of a population.

3. Monitoring performances and processes provides information which helps to empower right holders and enables them to hold duty bearers accountable for poor performance, unlawful conduct or wasteful use of public resources. Often right holders are not given data required to perform this task.

Recommendations:

1. Once right to food policies and programmes are in place, monitoring systems must be established with verifiable and time-bound targets and benchmarks to measure the progress made throughout implementation. Additionally and most importantly, there should be an institution with a specific mandate to monitor the respect, protection, and fulfillment of the right to food in the country.

2. The recommended system for monitoring right to food implementation is a human rights-based monitoring, meaning, it should measure both outcomes and processes and help bring out actual causal links of food insecurity.

3. Process indicators which capture participation, transparency and empowerment must be used in order to provide guidance on remedial action. Right holders must be given access to data in order to hold duty bearers accountable for their actions; that is, monitoring information should not be restricted to technical groups and decision makers, on the contrary, it should be available in a format and language that is accessible especially to the most food insecure groups.

Topic 5. Effective Action: Strategy and Coordination

Issues

The need for a human rights based strategy for the progressive realization of the right to adequate food in each country is explicitly recognized in the Right to Food Guideline 3.1: *'States...in consultation with relevant stakeholders and pursuant to their national laws, should consider adopting a national human rights based strategy for the progressive realization of the right to adequate food...as part of an overarching national development strategy'*. Such a strategy should serve to guide the mainstreaming of human rights principles and right to food approaches in overarching policy frameworks, such as national development or poverty reduction strategies, as well as in sectoral policies and plans. This approach emphasizes the linkage between the realization of the right to adequate food and socio-economic policies and programmes.

Gabriele Zanolli / FAO

A right to food strategy should address the four pillars of food security (food availability, access, stability, and utilization), describe policy and programme measures to be implemented by all sectors and, in particular, target the most vulnerable groups. The formulation of strategy should be guided by the outcomes of a comprehensive assessment of national legislation and policies, institutional and administrative frameworks, existing government programmes, and the current right to food situation. The strategy should address major constraints to the realization of the right to food and propose an agenda for change. It should also set clear time-bound and verifiable right to food targets and benchmarks.

The right to food strategy should be designed to promote awareness, empowerment and participation, all of which can help to sustain the political will necessary for the realization of the right to food. This strategy can provide a broad platform for the formulation of a food security and nutrition strategy because it defines public roles and responsibilities. It should indicate ways in which the government ensures accountability, provide for good coordination among civil society organizations and grass roots groups, and encourage broad-based participation in decision making.

In accordance with human rights principles, the process of developing the right to food strategy should be highly participatory and empowering, involving civil society organizations and grass roots groups; it should also be transparent with built-in accountability measures. Right to Food Guideline 3.5 encourages countries, individually or in cooperation with international organizations, to develop right to food strategies and incorporate the right to food into their overarching policy frameworks, such as poverty reduction strategies.

The formulation and implementation of multi-sectoral policy and programme measures requires inter-institutional coordination, so as to avoid duplication. Right to Food Guideline 5.2 encourages governments to establish national inter-sectoral coordination mechanisms to implement and monitor policies, plans and programmes, and to involve communities in the planning and implementation of government programmes. If countries entrust this coordination function to one institution only, that institution's mandate should be clearly defined, and regularly reviewed and monitored.

Challenges

National decision makers and policy planners often have a poor understanding of right to food concepts and principles, and of the importance of human rights in general. Moreover, national policy formulation teams lack the experience and knowledge to translate right to food principles into practical policy options or goals. An urgent and important challenge to be addressed for the formulation of a right to food strategy is, therefore, that of correcting lack of knowledge and experience. When doing so, the value added of incorporating right to food concepts and practices in food security and nutrition policies must be clearly expressed in practical terms.

While constitutions and policy preambles may declare a government's commitment to the right to food and to human rights in general, official statements and commitments are rarely translated into policy priorities, implementation strategies and actions. A right to food strategy should enable governments to give practical effect to their political will.

Due to its multi-sectoral nature, the food security mandate is the responsibility of a number of sectors and government institutions, none of which are recognized as being charged with the task or authority to lead and coordinate implementation. Even in situations where a multi-sector coordination body has been established with high level membership, it is often not granted the power to direct implementation, which severely limits its impact on sector plans and budgets. As has been pointed out by many of the panel participants, the institutional location of the coordinating body is also critical, and may determine the extent to which this body can mobilize and coordinate other institutions. A coordinating body that is housed in the president's or prime minister's office generally has greater authority than such a group located in a line ministry.

Inter-institutional coordination often meets resistance because: (a) it is perceived as interfering with the exclusive mandate of individual institutions; (b) it may lead to budget sharing among institutions, meaning, each institution may have to forfeit part of its budget; and/or (c) it is considered to create extra work without corresponding benefits or recognition for individual institutions. Cooperation between government officials and non-governmental or grass roots organizations is difficult where the latter's role is reduced to that of tracking government actions, or where governments see NGOs as competitors for international funding.

Country Experiences

In **Mozambique**, the PRSP 2006-2009 (PARPA II) includes food security as a cross-cutting issue and re-affirms the human right to food of all Mozambicans. It incorporates key human rights principles, such as equality, gender equity and non-discrimination, participation, transparency and accountability, the inherent dignity of all human beings, rule of law, and empowerment. The strategy paper highlights the need for a holistic approach to food and nutrition security, and recognizes health and social protection as human rights. After the adoption of PARPA II, Mozambique revised its first Food Security and Nutrition Strategy (Estratégia de Segurança

Alimentar e Nutricional – ESAN I), including a right to food assessment, and in 2007 adopted a revised strategy, ESAN II. The revised strategy's overall objective is to *'guarantee that all Mozambicans have at all times, physical and economic access to an adequate diet that is necessary in order to live an active and healthy life, thereby realizing their human right to adequate food.'*[42] ESAN II clearly defines the obligations of duty bearers and the rights of ordinary citizens. It calls for strong recourse mechanisms to allow the people of Mozambique to claim their right to food.

The Government of **Brazil** has made significant efforts to reverse the trend of food insecurity and hunger. Its most significant programme in the fight against hunger, known as *Fome Zero* (Zero Hunger), is a multi-sector and interdisciplinary approach to the realization of the right to adequate food. Programmes that form part of the strategy are divided into four categories: (a) increased physical and economic access to food; (b) promoting family agriculture; (c) income generation activities; and (d) social mobilization and education. Zero Hunger's most ambitious and broadest reaching programmes are the National Family Grant and the National School Feeding Programme. The National School Feeding Programme provided meals 200 days a year to 34.6 million school children up to 8th grade.[43]

The **Uganda** Food and Nutrition Council (UFNC) formulated the Uganda Food and Nutrition Strategy and Investment Plan (UFNSIP) as a follow-up to the 2003 National Food and Nutrition Policy (NFNP)[44]. The Strategy was endorsed by the Ministries of Agriculture and Health in November 2005, and awaits approval by Cabinet before it can be tabled in Parliament for debate and subsequent adoption. The Strategy's focus is on advocacy for good governance, inter-sectoral coordination, empowerment of duty bearers and rights holders, policy decentralization and gender equality. It aims to provide nutritional support to all women of child bearing age.

In the famous 'Right to Food Case' in **India**, the Supreme Court has made it the duty of every state and union territory to ensure that no one dies from starvation or malnutrition. This implies that people who are too poor to buy food should be guaranteed a minimum means of subsistence by the government, either through direct food aid or access to employment. The Court directed the states to ensure that all shops linked to the Public Distribution System are functioning. It also ordered the states to implement food-for-work programmes, a Mid-day Meal Scheme and the Integrated Child Development Services within a definite time-frame.[45]

There are some examples of national inter-sectoral coordination bodies that appear to be effective in coordinating and monitoring multi-sector activities, such as the National Council on Food and Nutrition Security (CONSEA) in **Brazil** and the Municipal Councils on Food and Nutrition known as COMAN in **Bolivia**. A Technical Secretariat for Food Security and Nutrition (SETSAN) has been

---

42   Mozambique. 2008. ESAN II. Chapter 3. 5.
     (Available at http://www.fao.org/righttofood/inaction/countrylist/Mozambique/Mozambique_ESAN_IIePASAN.pdf).

43   The Programme covers 47 million school children as of 2009.

44   http://www.pma.go.ug

45   The directions under various schemes include identification of beneficiaries as well as fixing the quantum of disbursement. For instance, the Court has directed state governments to implement the Midday Meal Scheme by providing every child in every government-run and government-assisted primary school a prepared meal of at least 300 calories and 8-12 grams of protein each day of school for a minimum of 200 days in a year. The order to implement this programme for all children in these schools makes it the largest school meal programme in the world, serving more than 50 million cooked meals daily.

established in **Mozambique**, while the Uganda Food and Nutrition Council (UFNC) has been created in **Uganda**. The effectiveness of such coordinating mechanisms seems to be related to their institutional location within the governmental hierarchy, as well as to their composition. For example, CONSEA reports directly to the President of Brazil; two-thirds of its members are representatives of civil society and one-third are government officials. The COMAN in Bolivia has some civil society members, but government officials are in the majority. These multi-sector coordinating bodies should be carefully assessed to see what else determines their effectiveness, and what their role should be in developing right to food strategies and mainstreaming the right to food in policies and programmes.

Conclusion

In order to ensure effectiveness throughout the process of implementation of right to food policies, it is necessary to begin with a strategy that has the right to food as its stated objective and is developed with broad participation of all stakeholders. Furthermore, human rights principles should be incorporated in order to turn identified right to food concepts into clear and practical policy options. The value added by right to food should be communicated to and understood by government institutions, NGOs and local communities in order to support cooperation of stakeholders. In addition, because the right to food requires action not only in the legislative branch of government but also in various ministries covering a variety of sectors, it is important to find effective means of coordinating and monitoring relevant intergovernmental activities – such as cross-sectoral planning and budgeting – to ensure effective action in fighting hunger.

Effective Action: Strategy and Coordination

Lessons Learned:

1. Strategies that help to translate principles and policy statements into policy priorities, action plans and practical implementation steps are best for advancing the right to food.

2. Coordination and cooperation among government agencies and other actors is crucial because very often multi-sectoral coordination bodies lack the power to guide implementation actions. Their placement in the government hierarchy plays an important role in terms of legitimacy, convening power and effective coordination.

3. Availability of financial resources is essential to implement policies and programmes. This is to say that when national budgets reflect right to food policy priorities and goals, there is a better chance of ensuring effective action towards increased food security.

Recommendations:

1. Strategies must identify the specific value added of the right to food in specific national contexts and address any lack of experience and knowledge on the subject of the various stakeholders.

2. Coordination bodies must be strengthened and should be hierarchically placed at the high ranks of government – such as under the direct control of a president or prime minister – in order to create a level of authority enough to direct State policy implementation.

3. From the very beginning right to food policies and programmes must be linked to annual budgeting processes – including sector budgets, mid-term expenditure frameworks and long term financial support guarantees. International donors should provide funding that is complementary to government funding especially when line ministries or governmental institutions are faced with insufficient resources at national level to move ahead with a right to food agenda.

III. FORUM CONCLUSIONS, RECOMMENDATIONS AND THE WAY FORWARD

The above summaries of the five panel discussions clearly indicate that the realization of the right to food is a long-term undertaking. It requires a change in mind-set implying a move away from the general perception that the food insecure are only seen as vulnerable people who need charity towards a recognition that they have rights and that support provided to them is the provision of entitlements which are inherent in every human being. Clearly, considerable time and resources will be needed to equip all members of society with a full understanding of the meaning of the right to food and how to put it into practice. It will also take a high level of commitment. The same can be said with regard to institutionalizing the right to food through establishing appropriate accountability and recourse mechanisms, promoting transparency in decision making and the use of resources, and creating political and social 'spaces' for true participation and the empowerment of the most vulnerable.

There has been progress in some countries, although a lot still needs to be done to achieve the full realization of the right to food for all. Countries that have made the right to food a national commitment and a government responsibility have, over the years, witnessed a steady and consistent drop in malnutrition rates, especially among the most marginalized groups, while institutional and policy framework have been strengthened to make these achievements irreversible and sustainable. The right to food will not happen overnight. It takes political will, consistency and time to lay the groundwork, formulate policies and implement laws.

The Right to Food Forum is an invaluable resource for people working on right to food and other related issues across the globe to meet and share their ideas and experiences. It is important for this type of exchange to continue. FAO can play an essential role in providing a continuous platform for such dialogue exchange, technical expertise, policy advice and capacity development.

FORUM CONCLUSIONS AND RECOMMENDATIONS

The Forum concluded with a clear message: the right to food is here to stay. The summary below does not aim at being comprehensive, nor does it reflect the richness of the debate during the panels and plenary sessions. Nonetheless, it remains more than useful to retain a brief synthesis from the large number of opinions expressed during the three day sessions. Readers will find a more detailed account provided by the Forum Rapporteur in the annex to this document. It is also available, with other rich documentation and contributions, at www.fao.org/righttofood

A) COUNTRY SUPPORT

The number of countries interested in strengthening the right to food and good governance principles in their policies, laws and programmes is rapidly increasing. Those that have made a first experience want to expand and intensify this work, especially as numerous guides and tools on how to put the right to food into practice are becoming available. There is a strong call for FAO to continue its work on the right to food and to increase its support to countries in their efforts to implement the Right to Food Guidelines at national level. Thus, the Forum

FORUM CONCLUSIONS AND RECOMMENDATIONS

marks not the end of a process, but rather the beginning of a new phase of implementation, with greater focus on country level activities, using the knowledge, tools, networks and strategies developed up to now. In this context, institutional capacity and training are particularly important, as well as improving monitoring and evaluation with indicators that are adapted to the country's situation. FAO confirmed its commitment to provide support for the implementation of the Right to Food Guidelines.

B) NATIONAL INITIATIVE AND COMMITMENT

Strong, well informed, national leadership on the right to food, particularly presidential leadership, can make a big difference, as has been demonstrated in some pioneer countries and reflected in the case studies presented in Part THREE of the present publication. Of equal importance is the fact that people must be informed of their human rights and empowered to request that these be respected.

C) PARTNERSHIP WITH CIVIL SOCIETY ORGANIZATIONS

Civil society organizations and other stakeholders are the main drivers of the right to food agenda, both at international and national levels. In many countries, they are critical partners supporting efficient governmental action in this area. In particular with regard to monitoring the implementation of the right to food, cooperation between government and civil society is of considerable importance.

D) STRENGTHENING THE RIGHT TO FOOD UNIT

The Forum demonstrated the leadership role of FAO and the Organization's convening power with respect to right to food work. The Right to Food Unit was a key element in supporting countries to make progress on this issue. Participants expressed concern about the future of the right to food work at FAO and asked for a strengthening of the Unit in the context of FAO reform. Countries will need the Unit's support to properly apply the guidelines on legislating, monitoring, assessing, budgeting and teaching the right to food. Moreover, there is a continuing need for an exchange of experiences on implementation at international level, at regular intervals. In this context, there was a strong call for a second forum on Right to Food.

E) INTEGRATION INTO FOOD SECURITY WORK

The Forum showed that it is essential to include the human rights dimension in food security work. The food crisis is an example of this. Safety nets and enhanced production are necessary, but not enough. The third track of the food security concept consists of the right to food related to good governance, that is, voice, participation, empowerment, non-discrimination, transparency, accountability, and rule of law. Policy makers dealing with the right to food and food security specialists should be brought together. FAO has an important convening power in this respect and should also provide a platform for inter-sectoral debate on right to food and food security issues at national level.

FORUM CONCLUSIONS AND RECOMMENDATIONS

F) INTER-DISCIPLINARY COMMUNICATION

One of the main difficulties encountered relates to communication. Politicians and economists do not always understand the language used by human rights advocates, and vice versa. It is of the utmost importance to translate the wealth of information into a language that is understood by the different groups, countries and sectors, without losing the essence of the message that is being conveyed.

G) SUPPORT TO HUMAN RIGHTS COMMISSION

Strengthening human rights commissions offers a great potential to promote the right to food. In addition to their traditional mandate to prepare reports to the Treaty Bodies, such institutions should monitor the national human rights situation with a view to achieving greater efficiency of governmental action. Their work on economic, social and cultural rights should be strengthened as well as their capacity to analyse the impact of the different policies and programmes on the human right to adequate food.

H) POLICY COHERENCE AT INTERNATIONAL AND NATIONAL LEVEL

Human rights have no borders. The hunger problem is not a series of national problems: it is everyone's problem. "A child might be born in a poor country, but that child is not in a poor world", a Forum participant said "A child is everyone's child and everyone everywhere has a responsibility towards that child." National and international strategies are interdependent and therefore it will be necessary to tackle the issues at both levels. However, national bodies are the primary duty bearers: the national or sub-national levels remain the place where duty bearers meet with right holders and where the practical efforts undertaken in support of the right to food will make a difference in people's lives.

I) IMPORTANCE OF HUMAN RIGHTS PRINCIPLES

While visible milestones have been reached in certain countries with respect to laws and political strategies, equal attention must be given to the less visible 'soft issues', such as empowerment, the strengthening of institutions, transparency, participation, non-discrimination and capacity building. These are more difficult to measure, but they are indispensable for the successful implementation of the more visible outputs in the form of laws, strategies, policies and programmes.

Such principles also mark the paradigm shift from charity to entitlements, from needs to rights. In Eleanor Roosevelt's words: *"Human rights is not something that somebody gives to you, it is something that nobody can take from you."*

Part THREE

COUNTRY CASE STUDIES

I. INTRODUCTION

Five country case studies were prepared for presentation at the Right to Food Forum. These studies describe the implementation of the right to food, discuss the achievements, challenges and opportunities encountered, as well as 'work in progress' in the following countries: Brazil, Guatemala, India, Mozambique and Uganda.

The case studies examine how these five countries have concretely promoted the right to food in the context of (i) identifying those most in need – the hungry and the poor; (ii) undertaking an assessment of the legal, policy and institutional environment and budgetary allocations, to indicate the needs of food security policy change; (iii) developing food security strategies; (iv) defining inter-institutional coordination mechanisms and the participation of non-governmental sectors; (v) integrating the right to food in national legislation (such as a constitution or framework law); (vi) implementing systems to monitor the impacts of policies and programmes, and other assessment measures; and (vii) establishing adequate recourse mechanisms (judicial, quasi-judicial and/or administrative) for those whose right to food has been violated. Capacity development is a cross-cutting activity in support of all the above implementation steps, and measures taken in this area are an integral part of the case studies.

As the work continues in these and other countries, the information will be updated. Unless otherwise indicated, the data contained in the texts below reflect situations presented at the Forum in October 2008 and updated where relevant to include major developments up to the end of December 2009.

II. BRAZIL
A Pioneer of the Right to Food

Main Points

The experience in Brazil has been one of the richest. Government and civil society working together have advanced the right to food on many fronts through effective laws, strong institutions, sound policies and an empowered civil society. This ensures that efforts to realize the right to food will continue as the country tackles the numerous challenges ahead.

- The multisectoral and participatory anti-hunger programme Fome Zero (Zero Hunger) was created by the President of Brazil in 2003 to focus on the country's 11 million poor families.
- It has reduced malnutrition and improved the eating habits of these poor families. In fact, Brazil has achieved the MDG of reducing extreme poverty and hunger by half. Since 2003, 14 million Brazilians have been taken out of extreme poverty.
- The school feeding programme currently provides 47 million students with a nutritionally-balanced mid-day meal each day. It also teaches the importance of proper nutrition as well as basic facts regarding food and the necessary components required for a healthy diet. The programme is now based on a school feeding law (adopted in 2009) that has as its stated objective the realization of the right to food.
- The National Council on Food and Nutrition Security (CONSEA) was re-established in 2003 as an advisory body to the President of Brazil, to develop policies and guidelines for guaranteeing the right to food. In 2005, it created the Standing Commission on the Human Right to Adequate Food to examine public programmes and policies. CONSEA has been effective in furthering the implementation of the right to food and encouraging civil society participation in the development and implementation of right to food policies and programmes.

1. Background

One-third of Brazil's export earnings come from agriculture, and yet a study undertaken in 2004 showed that 72 million of its 185 million inhabitants were affected by food insecurity.[46] Food-related demonstrations occurred in the streets while various other manifestations and actions at grassroots level highlighted the fact that despite economic growth, too many Brazilians were living on the brink of starvation.

Luis Inácio Lula da Silva, a former union leader, was elected President in 2002 amid promises to end hunger in Brazil. In his inaugural address in 2003, he stated: *"In a country with such fertile soil and so many people willing to work, there is no reason to let hunger exist. ... among the priorities of my government should be a programme of food and nutrition security named Fome Zero (Zero Hunger)."* He went on to assert, *"As long as one of our Brazilian brothers is hungry, we have enough reason to be embarrassed."*

---

46 Brazilian Institute of Geography and Statistics. 2004. *Segurança Alimentar.* Pesquisa nacional por amostra de domicílios (PNAD 2004), p. 148. Rio de Janeiro. Publicação conjunta MPOG/FIBGE e MDS.

Zero Hunger has made progress, although it is slow progress for those living in the grips of hunger and poverty, and the Brazilian Ministry of Social Development and Fight Against Hunger is striving to improve implementation. The Zero Hunger programme is attempting to confront the problem by:

- creating awareness and commitment at the highest political level;
- generating and extending widespread support to the poorest people in the country;
- taking action through agriculture and production initiatives, and also across all sectors of government; and
- forging a partnership between government and civil society for a rights-based, participatory process.

Brazil has implemented numerous programmes and initiatives that conform to this human rights based process and has taken many of the steps recommended in the Right to Food Guidelines adopted unanimously by FAO. In fact, many of the country's experiences were being widely discussed while the Guidelines were under negotiation.

2. Brazil's Political and Agricultural Past

Brazil's hunger problem has deep roots. In the sixteenth century, Brazil was a Portuguese colony with a thriving agricultural economy thanks to slave labour; indigenous people were largely ignored. In 1822, the son of the King of Portugal created a constitutional monarchy in Brazil and declared himself Emperor. In 1900, the country became a federal republic and outlawed slavery.

Over the next century, Brazil's indigenous population and the descendents of former slaves known as *quilombolas* continued to be marginalized and remained subsistence farmers or worked as low-paid labourers in fields and factories if and when work was available for them.

The government was overthrown in 1930 and this led to half a century of political and social upheaval. It was the beginning of Brazil's industrialization, however. It prompted a surge in social movements, trade unions and a drive for land reform. All of this came to a stop with the coup d'état of 1964. The new leadership adopted policies that reinforced the agricultural export model, leading to an enormous concentration of land property and increased rural and urban poverty.

In the 1980s, the military leaders relaxed their hold on the country and Brazil began moving towards democracy. Political reorganization was crystallized in the Constitution of 1988, which brought human rights to the forefront. A direct presidential vote the following year heralded Brazil's return to a popularly elected government.

However, the 1990s brought foreign debt crises and rampant inflation, leading to widespread social pressure. The organization known as 'Citizenship Action Against Hunger, Extreme Poverty and for Life' united half of the population on the need to discuss hunger issues.

All this turmoil provoked some change. Government statisticians produced a Hunger Map[47] indicating, region by region, that 32 million Brazilians were living in extreme poverty. A Plan Against Hunger and Extreme Poverty was drafted and firmly rooted on partnership, solidarity

---

47   Instituto de Pesquisa Económica Aplicada. 1993. *O Mapa da fome: subsídios à formulação de uma política de segurança alimentar.* Anna Maria T. Medeiros Peliano(coord.). Brasília.

and decentralization. The National Council on Food and Nutrition Security (CONSEA) was established and the nation's first National Food and Nutrition Security Conference held in 1994. Attendance included broad civil society representation as well as that of government officials.

The best of government intentions and ambitious projects were waylaid by international pressure to reduce State spending. Small industries and landholders were hit particularly hard. As poverty and malnutrition worsened, social pressure continued to build.

In 1992 Brazil ratified the ICESCR. A Special Secretariat for Human Rights was set up in 1997 and a National Plan for Human Rights was developed in 1999 and 2002, but changes were not happening quickly enough. Years of intense social and economic problems, as well as the impeachment and resignation of President Fernando Collor in 1992, created the impetus for inevitable change in Brazil.

Criminal activity, especially in the cities, increased rapidly in an environment of desperation and inequality. There were calls from the streets to address the country's gross inequities. All of this made the working class figure of Luiz Inácio Lula da Silva an appealing candidate in the presidential election in 2002.

Once elected, the President vowed to end hunger in Brazil. Zero Hunger, the commitment to eradicate food and nutrition insecurity, has become a national goal embraced by Brazil's Government and civil society alike. It is an ambitious effort and has received international support and attention. Zero Hunger is endorsed by experts from FAO, the World Bank and the Inter-American Development Bank and receives technical support from FAO for different projects.

The anti-hunger programme, which includes both direct assistance and long-term poverty alleviation measures, requires coordinated action by all areas of government at federal, state and municipal levels. Extensive participation of all segments of society continues to be essential.

Brazil has made significant progress but a lot more is needed if its new development model of social inclusion is to be achieved. The country continues to rank among the most unequal in the world and the concentration of land ownership in the hands of a minority continues to be a concern for the landless. Large economic discrepancies among regions and ethnic groups are a constraint to sustainable development.

The efforts made so far must be continued if the progressive realization of human rights, including the right to food, is to be successfully achieved.

3. Identifying the Hungry in Brazil

While famine is infrequent, millions face chronic food insecurity. Many are *quilombolas*, indigenous people or migrant groups, all of which have been largely ignored.

In 2004, the Brazilian Institute of Geography and Statistics known as *Instituto Brasiliero de Geografia e Estatistica* (IBGE)[48] undertook a food security survey to determine whether Brazilians had enough to eat and whether their access to food was limited or sporadic. While it did not

---

48   Brazilian Institute of Geography and Statistics. 2004. *Segurança Alimentar.* Pesquisa nacional por amostra de domicílios (PNAD 2004), p. 148. Rio de Janeiro. Publicação conjunta MPOG/FIBGE e MDS.

measure the quality or nutritional content of the food, it did measure the psychological strain that accompanies one's inability to meet family needs. The survey showed areas where the right to food was not being realized, and established Brazil's baseline for measuring the impact of public policies and assessing progress in fighting hunger.

According to this survey, food insecurity affects 72 million Brazilians (18 million households) or 39.8 percent of the population. Those slightly insecure are 32.6 million or 18 percent; the moderately insecure are 25.5 million or 14.1 percent; and the severely insecure are 13.9 million or 7.7 percent.

Additional data, gathered by health teams during vaccination campaigns and from the Indigenous Health Group of the Brazilian Health Foundation, indicated the need for the Ministry of Health to prioritize the promotion of food security and nutrition, especially among the country's most vulnerable groups. Official statistics had traditionally overlooked Afro-descendents, indigenous populations and other particularly vulnerable groups, such as the populations living in remote geographic regions. In 2006, the Ministry of Social Development and Fight Against Hunger gave special attention to the living conditions of the *quilombolas*.[49] Some of the results are indicated below:

Quilombola Families benefiting from at least one social programme (sample = 2155)	%
Bolsa Família or National School Meal Programme	51.7
Food supplement programme	6.5
Installation of equipment	1.6
Prevention of child labour (*PETI*)	3.8
Social assistance (*Benefício de Prestação Continuada*)	2.3
Rainwater cisterns project (*Projeto Cisternas*)	3.2
Food Procurement Programme (*Programa de Aquisição de Alimentos*)	8.0
Others	5.0

The surveys and medical examinations generated the most comprehensive population profile Brazil had ever produced. Disaggregated data enabled profiling by region, ethnicity, race, gender and age, and facilitated the identification and mapping of specific populations and regions most vulnerable to food and nutrition insecurity. This information is crucial in order to address the particular dietary needs of various groups.

Disaggregated data together with Brazil's Unified Register assist in the provision of support to those in need. Any family that earns less than one-half of Brazil's defined minimum monthly salary

---

49   Brasil. 2006. *Chamada Nutricional "Quilombola"*, Ministério do Desenvolvimento Social e Combate à Fome.

(US$192.06)[50] is enrolled on the register. If each family member earns less than US$66 on average, the household is eligible for the Family Grant programme which was instituted in 2004 as part of the Zero Hunger Programme. Once registered, beneficiaries can apply more easily for Zero Hunger's 52 programmes.

Pilot Projects

In 2005, an NGO named Brazilian Action for Nutrition and Human Rights (ABRANDH), together with Brazil's National Right to Food Rapporteur, examined hunger and poverty situations in two slums. The pilot projects in (i) Vila Santo Afonso, Teresina City (State of Piauí), and (ii) Sururú de Capote, Maceió City (State of Alagoas), found hundreds of families living in housing built from mud or plastic sheeting, with sanitation that was far from hygienic.

For years Vila Santo Afonso has been home to a multi-ethnic, fluctuating population, with a core group of some 250 families. Living in what Teresina city defined as an 'irregular area', these families were considered squatters. A health centre in Teresina even posted a warning that Vila Santo Afonso residents were not eligible to receive treatment there, despite living in the area of coverage. Sururú de Capote, with 450 families, revealed a similar situation.

As part of the project evaluation exercise, children under five years of age underwent medical examination. In Sururú de Capote, more than 80 percent were found to be anaemic and 87 percent tested positive for intestinal parasites. Nearly half of the children examined were below the average height for their age and most were underweight. In Vila Santo Afonso, severe food insecurity affected 54 percent of the households – almost eight times the national average. Families attempted to send their children to school but the area lacked classrooms. While most people reported that they had meals every day, their diet sometimes consisted of rice only, which is hardly nutritionally adequate, and there was usually an insufficient quantity for the entire family.

With this information in hand, NGO workers and community leaders were able to demonstrate to Teresina officials that many families were indeed eligible for Brazil's Family Grant (Bolsa Familia) Programme, in spite of the fact that they had previously been denied benefits. In addition to their 'irregular' slum addresses, many family members also lacked birth certificates or other suitable identification.

With Teresina city officials accepting Vila Santo Afonso homes as legitimate residential addresses, families there could now apply for national assistance. Brick and mortar houses were then built and the whole community was included in the Food Procurement Programme. As *bona fide* residents, the families have also become eligible to avail of the city's medical services. By increasing public classroom capacity, more children will be able to receive government lunches every day when they go to school.

---

50   According to the exchange rate of 3 March 2009 1US$ = R$ 2.421 per Central Bank of Brazil (*Banco Central do Brasil*).

Pilot Projects

ABRANDH supported community initiatives to: a) establish partnerships with different civil society organizations; b) participate in public forums that facilitated monitoring public policies such as Public Policy Councils; c) demand the implementation and adjustment of public policies by public entities; and d) complain to rights protection institutions, such as the Public Prosecutors and Human Rights Councils, when their demands were not met by the executive branch (further details are provided under the section entitled 'Legal and Administrative Recourse Mechanisms').

Progress is being achieved slowly but steadily for those whose rights are most at risk. Thanks to the Unified Register, both the MDS and relevant government officials at all levels have better information to identify the poor and their whereabouts, and can thus target these vulnerable populations with better and more efficient hunger-reduction strategies. Work towards realizing the right to adequate food has also led to improvements in education and housing. The adoption of a human rights based, participatory approach has paid off for local residents by increasing their dignity and self-respect.

4. Assessing Laws, Policies and Institutions

The Right to Food Guidelines encourage States to assess their laws, policies and institutions so as to identify issues hindering the right to food. A careful assessment will reveal the current situation and a conscientious analysis will point to the necessary policy changes and measures required to improve the food security situation of the populations concerned.

While acknowledging Zero Hunger's efforts to eradicate hunger, CONSEA noted that the programme lacked some essential elements needed to make it a rights-based approach to realizing the right to food. In 2005, it established the Standing Commission on the Human Right to Adequate Food to advise the government on how to incorporate the right to food in public policies.

Through adapting an assessment tool developed by South Africa's Human Rights Commission, Brazil's Standing Commission looked at Zero Hunger's most significant social assistance programmes:

(1) National School Feeding (Ministry of Education) has provided some 200 meals a year to 34.6 million school children up to 8th grade – a total of nearly 6.9 billion meals. According to the latest update from the Ministry, as of 2010 the programme has been extended and is currently providing meals for 47 million students at a cost of approximately 3 billion Reais.[51]

(2) National Family Grant (Ministry of Social Development and Fight Against Hunger), which is an income transfer programme, reaches 15.7 million families with an average monthly stipend of US$ 49.56; and

(3) National Family Health Strategy (Ministry of Health) strengthens national healthcare, with 29 300 family health teams covering 49.5 percent of the Brazilian population (93 million people).

---

51 http://www.fnde.gov.br/index.php/programas-alimentacao-escolar

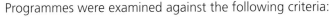

Programmes were examined against the following criteria:

- clear definition of right holders and duty bearers;
- empowerment and informed participation of right holders;
- accountability of the State to respect, protect and fulfil the right to food;
- creation and strengthening of claim mechanisms;
- defining and monitoring of goals, benchmarks and deadlines;
- strategies for information dissemination; and
- capacity building for both duty bearers and right holders.

CONSEA's Standing Commission on the Human Right to Adequate Food assessed the Family Grant Programme, Zero Hunger's most important effort, against the above criteria. It reaffirmed the importance of family grants for realizing the right to food of Brazil's marginalized populations, and recommended strengthening the Programme with social control mechanisms. These would scrutinize applicants more accurately, ensuring that all eligible families are enrolled, and that those who are ineligible are not included. The Commission also recommended that MDS, together with Brazil's Public Prosecutors, design and implement accessible recourse mechanisms through which right holders could appeal to claim their rights.

The school feeding assessment led to similar recommendations so that the vitally important programme could be strengthened. School Feeding Programme officials worked with the Standing Commission to create tools through which right holders could claim their right to food. Informing school children about their rights empowers them to notify someone if for some reason their rights are not being met. This mechanism raises children's understanding of rights and helps create an enduring culture of human rights.

In March 2007, after further assessment, CONSEA advised the Ministry of Education to amend the school feeding law to take into account Brazil's cultural diversity and, in particular, the nutritional deficiencies of certain vulnerable groups. It also recommended that the school feeding programme link more closely with another government programme that supports smallholder farmers by purchasing food locally.

In addition to programme specifics, analyses revealed other information that may be useful in subsequent policy assessments. For example, interviewers found that public implementing agents were far more receptive to changing policies than were higher political authorities although some Brazilian programmes and policies already incorporate human rights dialogue, few had mechanisms in place through which those rights could be demanded.

Although law and public policy express the right to food, its realization depends on a number of interdependent prerequisites. These include stakeholder awareness, capacity and participation, independent human rights institutions, effective and accessible claim mechanisms, and civil society mobilization.

It is important to assess programmes, laws and institutions in order to identify challenges to the realization of the right to food and decide what measures are required. The method used in Brazil for a human rights based assessment of public policies includes simple steps that can be adapted for use by institutions and commissions at all levels of government. They can also be adapted for use in other countries.

5. Sound Food Security Strategy

In order to fight hunger effectively, short term relief must be coupled with long range improvements. While mutually reinforcing, the former creates opportunities and the latter equips the hungry to take advantage of these opportunities. Countries that follow this 'twin-track' approach towards food security achieve faster and better results in their fight against hunger.

The Right to Food Guidelines recognize that mechanisms to ensure that duty bearers meet their obligations, and provisions to enable rights holders to claim their rights, have to be implemented progressively. Consequently, CONSEA's third national conference in 2007 closed with a final declaration stating that the guiding principles for food security policy should 'incorporate principles and mechanisms for claiming the right to food so as to eliminate dependency-forming practices that are charity-based, rewarded or traded, in order to promote a culture of rights.' This is still work in progress.

It takes time to empower disadvantaged people to take charge of their own destiny and further measures are required to achieve this. The country is committed to allocating resources for social safety nets while building towards more permanent solutions.

A sound food security strategy is a roadmap, discussed and agreed upon by all, in which there is government-coordinated action towards a common goal. This includes developing hunger and malnutrition eradication strategies with targets, time frames, clearly allocated responsibilities and evaluation indicators that are known to all.

Zero Hunger's 52 programmes[52] come under four thematic areas: increased physical and economic access to food; promotion of family agriculture; income generation activities; and social mobilization and education. By adding social mobilization to the 'twin-track' formula, Brazil enhanced the returns of Zero Hunger, increasing participation to include not only the poor but also any member of the community who wishes to participate.

Under the political leadership of the President and his cabinet, Brazil allocated US$62.4 billion[53] for this programme between 2003 and 2008. Most of the operations are coordinated by the Zero Hunger Working Group led by MDS. Zero Hunger's most ambitious, broadest reaching programme has been Bolsa Familia, or the Family Grant. This programme, under the MDS, makes monthly income transfers to some 15.7 million poor families, provided that their children attend school and participate in preventive health monitoring. The Ministry of Education's National School Feeding Programme, one of the world's largest programmes of this type, provides at least one daily meal for school children and adolescents and has reached more than 47 million children.

Not all development programmes are funded exclusively by the central government. Throughout Brazil's arid northeast area, federal, state and municipal governments partnered to build 219 000 cisterns and they aim to build an additional 800 000. Small-scale irrigation systems helps more than 34 000 farmers to earn an income, while the availability of fresh drinking water gives communities a healthy safety net. Private and corporate donations have helped to fill and distribute almost five million food baskets for vulnerable populations and 91 food banks have been set up in 19 of Brazil's 26 states.

---

52   http://www.mds.gov.br/fomezero
53   http://www.planalto.gov.br/consea

Seventeen states host nearly 400 community gardens and more than 300 000 urban families have received assistance to finance small garden plots for their own consumption needs and for income generation. In cities with more than 100 000 inhabitants, over 100 canteens (cafeterias) serve nutritious US$0.50 meals to approximately 2 000 low-income workers (per canteen). These canteens purchase supplies from local farmers and provide on-the-job training opportunities for kitchen workers.

Income generation schemes may be time-consuming to implement but they are essential in order to sustain the right to food and food security. Nearly 400 000 people received skills training and US$1 billion have been provided in micro-credit loans since 2003. Brazil's unemployment level, however, remains at nearly 10 percent, and 31 percent of the population remain below the poverty line.

Brazil's battles against hunger have had a rapid success in many areas. According to IBGE data, poverty decreased by 8 percent between 2003 and 2004.[54] Zero Hunger is credited with reducing inequality to the lowest level Brazil has seen in more than 30 years. The country is right on track but its goal to end poverty, rooted in centuries of inequality and marginalization, will take much longer to be fully realized.

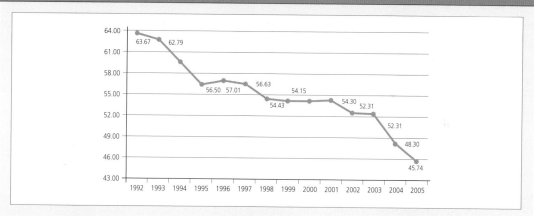

Reduction in Rural Poverty between 1992 and 2005
(percentage of individuals with monthly income lower than R$125)

Source: CPS./IBRE/FGV processing microdata from PNAD/IBGE

---

54   Brazilian Institute of Geography and Statistics. 2006. *Pesquisa Nacional por amostra de domicílio: Segurança Alimentar 2004*. Rio de Janeiro. IBGE. (Available at www.ibge.gov.br).

6. Allocating Roles and Responsibilities

Allocating clear roles and responsibilities to different government sectors and levels is essential for ensuring transparency and accountability. Although international law defines the State as the primary duty bearer for realizing the right to food, specific obligations and responsibilities may be delegated to various institutions. The Brazilian Government has assigned specific obligations to many state agencies but all acknowledge that civil society participation is essential to the process.

While Brazilian President Lula is the political leader and overall motivator for the realization of the right to food, MDS is directly responsible for government efforts to relieve immediate needs and address the underlying causes of hunger and food insecurity. In addition to implementing the Family Grant cash transfers, the MDS oversees all 31 Zero Hunger programmes including the Ministry of Education's National School Feeding Programme, the Ministry of Agrarian Development's Programme for Family Agriculture and the Ministry of Health's Malnutrition Control and Prevention Programme. Furthermore, the MDS chairs and convenes the Inter-ministerial Chamber, established by the food security framework law (described in the following pages under Legal Framework to Realize the Right to Food).

Brazil's National Council on Food and Nutrition Security (CONSEA)

- CONSEA's ground-breaking discussion of social exclusion, although short-lived (1993-94), turned Brazil's hunger problem into its key political issue. Citizens mobilized in a way that the country had never witnessed before and demanded to participate in changing public policies.
- Re-established in 2003, CONSEA's national counsellors – 38 from civil society, 19 from the government and 16 non-voting observers – are the President's consultants and advisors on food security and right to food issues. Hence, non-governmental organizations hold a 2/3 majority in CONSEA. Sector groups (such as production and supply, nutrition and health) and commissions (such as indigenous or black populations and right to food) contribute to monthly plenary agendas. Recommendations are then forwarded to the President.[10]
- Decentralized CONSEAs have also been formed in all 27 Brazilian states and even in some municipalities. The National Food and Nutrition Security Conference brings together 1300 counsellors from around the country for a national assembly which, by law, must convene at least once every four years. These delegates, together with approximately 500 non-voting attendees, make up the main body of CONSEA at the national level, responsible for monitoring and establishing guidelines for the implementation of the National Policy on Food and Nutrition Security.

---

55   Decree No. 6.272 (23 November 2007) defines the competence, composition and functioning of CONSEA.

State-level Public Prosecutors have a constitutional mandate to promote and defend human rights. Their Working Group on the Right to Adequate Food has published and distributed a handbook that helps prosecutors handle right to food violations. Prosecutors in Maceió and Piauí have legally charged officials for failure to meet their obligations.

An autonomous human rights agency is essential for monitoring human rights and for reporting violations. Brazil's Special Secretariat for Human Rights maintains the status of a ministry. When pilfered funds jeopardized local school feeding programmes, its Special Commission for Monitoring Violations of the Human Right to Adequate Food, together with the Ministry of Education and other government agencies, discussed alternatives to suspending the programme. A working group was set up to propose a new law which would ensure the continued distribution of school meals and thus avoid school children from being penalized as a result of corruption. The report presented by this group provided the basis for the new school feeding law.

The independent National Rapporteur for the Right to Food and Rural Land, installed by a civil society network in 2002, has the task of investigating right to food violation claims, increasing the visibility of these cases and working towards remediation. Investigations may lead to public hearings and recommendations made to public authorities. The Rapporteur teams up with Public Prosecutors, the Special Secretariat for Human Rights and others, to monitor fulfilment of the recommendations.

More than 100 NGOs, social movements, networks, researchers and activists formed the National Forum on Food and Nutrition Security while preparing for the 1996 World Food Summit. Continued pressure by civil society is keeping food security and nutrition on the political agenda. The training strengthens participation and improves the monitoring of right to food violations.

While Brazil's right to food momentum came from an internal surge of political and social enthusiasm, assistance from the international community was welcomed by all national stakeholders. FAO was one of Zero Hunger's first evaluators and the FAO Right to Food Unit continues to cooperate with the government and civil society organizations.

Brazil's experience shows that realizing the right to food is not a quick and simple process, even with all stakeholders working together. But without such stakeholder cooperation, realization would be impossible.

7. Legal Framework to Realize the Right to Food

Following the recommendation by the second National Food and Nutrition Security Conference in 2004 to draft a framework law, the Government of Brazil assigned this task to CONSEA's Food and Nutrition Security System Working Group. The draft law was sent to the Brazilian Parliament in October 2005 and given top priority. In September 2006, after less than a year of negotiation, Parliament approved Brazil's National Food and Nutrition Security Framework Law (LOSAN)[56] and the President signed it into law.

---

56 Law No. 11.346, *Cria o Sistema Nacional de Segurança Alimentar e Nutricional – SISAN com vistas em assegurar o direito humano à alimentação adequada e dá outras providências* (English translation: Creates the National Food and Nutrition Security System – SISAN, to guarantee the human right to adequate food and takes other measures), Presidency of the Republic, Civil Cabinet, Subcommand for Juridical Affairs, 15 September 2006.

The key provisions of LOSAN are as follows:

- Adequate food is a basic human right, inherent to human dignity and indispensable for the realization of the rights established by the Federal Constitution. The government shall adopt the policies and actions needed to promote and guarantee food and nutrition security for the population.
- The government shall respect, protect, promote, provide, inform, monitor, supervise and evaluate the realization of the human right to adequate food, as well as guarantee the institution of specific claim and recourse mechanisms.
- The national food and nutrition security system seeks to formulate and implement policies and plans on food and nutrition security, motivate the integration of efforts between the government and civil society, as well as promote the examination, monitoring, and evaluation of Brazil's food and nutrition security.

The Inter-Ministerial Chamber in charge of coordinating the Food and Nutrition Security System needs to elaborate the National Food and Nutrition Security Policy and Plan, "suggesting the guidelines, goals, basis of resources and tools to examine, monitor, and evaluate its implementation."[57]

Brazil's Constitution and other laws spell out social rights, such as the right to land, health and housing. However, LOSAN became the first law in Brazil that focuses specifically on 'how to' realize an economic, social and cultural human right. The law even goes beyond incorporating the human right to adequate food concepts, demanding that the government create mechanisms for monitoring and evaluating implementation progress, although subsequent legislation will be needed to clarify details.

Cooperation among government and civil society groups drafting LOSAN produced a balanced, functional law. This landmark team project also demonstrated that a human rights based approach, which includes as many stakeholders as possible, enhances understanding and appreciation, as well as building the capacity of all participants.

LOSAN has woven Brazil's collection of *ad hoc* right to food and food security systems into the fabric of the country's institutional structure. It ensured that CONSEA and the National Food and Nutrition Security Conference – the country's main hunger eradication coordination entities – became permanent parts of Brazil's food security governance after a series of decrees were adopted to regulate the specific provisions of LOSAN. With the ending of hunger now raised to a permanent State objective, it is no longer subject to changing governments or new presidents, as was the case in 1994 when the first CONSEA was abolished.

8. Monitoring the Right to Food

Brazil began with a careful analysis of its starting point. Ten-year-old census data was updated in 2002-03 through a nutrition, food spending and family food security survey carried out by the Brazilian Institute of Geography and Statistics.

---

57   Decree No. 6.273 (23 November 2007) created the Interministerial Chamber for Food and Nutrition Security (*Câmara Interministerial de Segurança Alimentar e Nutricional*) and the National Food and Nutrition Security System (*Sistema Nacional de Segurança Alimentar e Nutricional*).

Brazil already had considerable experience in monitoring the food security situation, yet the government asked for CONSEA's help in incorporating the human rights dimension in monitoring. During the third National Food and Nutrition Security Conference, CONSEA was given the task of going still further by making '... *a broad and critical analysis of the food security and nutrition situation of the country, guided by the right to food ...*' It was also asked to recommend specific actions to '... *monitor the implementation of the Right to Food Guidelines ...*'[58]

The CONSEA Working Group for Indicators and Monitoring suggested that Brazil begin by using existing indicators produced by various Government ministries, such as Health, Education, Agriculture and Social Development. The Working Group's proposal includes 26 indicators in seven dimensions:

- Food production
- Availability
- Income
- Access
- Health and access to health services
- Education
- Public policies that promote food security

Specific indicators include the percentage of family income spent on food and the price variations of a food basket containing Brazil's customary food staples. Another indicator is the total national production quantity in terms of calories per capita.

Compliance with the Right to Food Guidelines entails the progressive realization of the right to food to the extent of available national resources. Transparency is essential in order to monitor whether or not the Government is spending wisely or complying with the needs of the population. For the past few years, Brazil has been placing its annual budget (http://www.mds.gov.br/transparencia) on the web, as well as the monthly expenditure reports of MDS (http://www.mds.gov.br/mds-em-numeros).

As the Government planned its pluri-annual budget for 2008-11, civil society was represented by various councils, such as CONSEA. The third National Food and Nutrition Security Conference in 2007 recommended that:

- Budgeting should consider food security and nutrition concerns;
- Budget allocations at all levels of government should finance the food security system; and
- A line item should be included, containing all food security programmes.

CONSEA began working on the budget in 2003, and has been sending advisory briefs to the President, together with recommendations and spending priorities. Recently, Brief n° 008/2008 requested a budget for several new activities, including expanding school feeding for older students, right to food education for healthcare personnel and the construction of additional cisterns.

---

58   CONSEA. 2007. III National Food and Nutriton Security Conference *(III Conferência Nacional de Segurança Alimentar e Nutricional)*. Final Report. Brasília.

When monitoring whether or not the right to food is realized, any violation of this right must also be included. While the MDS is concerned with realizing the right, the Secretariat for Human Rights, Public Prosecutors and Brazil's National Rapporteur deal with violations. Between 2004 and 2006 the Rapporteur conducted 19 missions to examine right to food violations. He and his team investigated human rights issues facing the country's homeless, landless rural workers and many urban marginalized populations. While working in 11 Brazilian states, the Rapporteur also took the opportunity to distribute educational and information materials.

The more duty bearers and right holders learn about the right to food, the better they can participate in the monitoring process. States, in turn, can improve the efficiency of these agencies not only by delegating responsibility but also by training those assigned and providing them with the necessary resources to perform the job. Failure to monitor progress is an invitation for any programme to stray off track, whereas nothing ensures better monitoring than that of enlisting every Brazilian in the country to assist in the effort.

9. Legal and Administrative Recourse Mechanisms

Independence and autonomy enables Brazil's National Rapporteur on the Right to Food, Water and Rural Land not only to investigate claims brought by individuals, groups and communities but also to support the creation of institutional recourse instruments. Rooted in the social movement and fully supported by Brazil's Public Prosecutors and international human rights organizations, the Rapporteur has become the voice of the many unheard or neglected groups and issues.

Following the completion of an investigation, both the Rapporteur and the Federal Public Prosecutors hold public hearings with claiming parties, the civil society groups involved and the public officials responsible for the situation. Not all hearings result in the correction of rights violations but they all bring right to food issues to the attention of government authorities, the media and the general public.

The Special Commission for Monitoring Violations of the Human Right to Adequate Food, under Brazil's Special Secretariat for Human Rights, is the country's first commission to specifically address an economic, social and cultural right. Just as is the case with the National Rapporteur, this Special Commission investigates violation claims and proposes reparation. When school feeding programmes were jeopardized by pilfered funds, the Special Commission, along with the Ministry of Education and other government agencies, discussed alternatives. The suspension of fund transfers would have deprived the children of their lunch entitlement so the group suggested several alternatives including working with Public Prosecutors, publishing identities ('naming and shaming') of guilty parties and enlisting local school councils to take part in overseeing spending. Another non-judicial remedy was to simply publish the budgets allocated to all public schools. This enabled parents, teachers and even students to demand that school officials account for the meals being served.

Brazil's constitution grants any individual or group the Right to Petition, giving them the right to notify (in writing) public authorities of violations and to demand reparation. Authorities receiving such petitions are obliged to act or to at least redirect claims to the appropriate bodies tasked with handling specific violations. Programmes such as the Family Grant could benefit from the type of redress instruments already built into Brazil's *Benefício de Prestação Continuada*.

These mechanisms enable the elderly poor, whose benefit applications may have been rejected, to file a legal claim for a human rights violation.

Without investigation by civil society organizations that report on and monitor human rights violations, many cases might continue to be ignored. In cooperation with the Rapporteur, the Public Prosecutor or other agencies and civil society organizations have reported violations confronting migrant rural workers, indigenous people, *quilombolas*, and populations evicted by the construction of dams or by urban development.

Autonomy from the executive, legislative and judicial branches enables the Public Prosecutor to represent civil society before these three branches and to hold all government bodies responsible for enforcing the law. The Public Prosecutor at any level (federal, state, military or labour) can convene a public hearing to investigate individual or group claims of rights violations and can facilitate the adoption of remediation measures.

When a hearing involves conflict between the State and society, the Public Prosecutor has the authority to summon government bodies and other relevant parties, and negotiate a Terms of Conduct Adjustment. This is a legally binding settlement that defines the obligatory actions of all parties and establishes a time-frame within which the identified violations must be corrected. Public civil suit is considered to be a last resort.

In one case, a state-level Public Prosecutor, concerned about conditions in Sururú de Capote, appealed to the City of Maceio, State of Alagoas, for a public civil action. In 2007, on the basis of identification of human rights violations, the Public Prosecutor of Alagoas filed a public civil action[59] to demand children's and adolescents' rights in the Sururu de Capote community. This was the first public civil action in the country demanding the realization of the right to food and the human rights of children and adolescents to education, life and health, based on international human rights laws and on domestic legislation. The action was granted by the judicial branch of Alagoas. This has set a precedent that can serve as a basis for other actions and judicial decisions on behalf of the economic, social and cultural rights of vulnerable communities.

The case was decided in favour of the community in September 2007. According to the court order, the Municipality of Maceió had to take the following measures:

- Set up a multidisciplinary commission to analyze the socio-economic profile of children and adolescents living in the slum within 30 days;
- Provide adequate conditions for the city's Guardianship Council to operate within 30 days;
- Ensure that there is enough shelter for children of less than 18 years within 30 days;
- Offer day nurseries for infants of less than 6 years within 30 days;
- Ensure the enrolment of all children and adolescents at primary school age within 30 days;
- Propose short, medium and long-term solutions to the community within 90 days;

---

59 Public Civil Action is the procedural instrument based on alleged damage or threat of damage to consumers, the environment, the urban order, the economic order or any other diffuse or collective interest, i.e., the interest not of an individual but of groups or even the entire society. The Public Ministry of Alagoas has filed a Public Civil Action demanding measures that could guarantee the human right to adequate food and other economic, social and cultural rights of communities living in situations of extreme poverty in Maceió. The final version of Public Civil Action is available on ABRANDH's website (http://www.abrandh.org.br). The number of the process generated by the public civil action is PROC. N° 4.830/07.

- Ensure that sufficient resources are allocated to these solutions in the 2008 budget and prepare a contingency plan if enough resources cannot be found;
- Expedite the registration of children and adults;
- Start a campaign against child prostitution.

The Municipality of Maceió appealed against the Court's decision. While a final settlement is still awaited, the court case can contribute to awareness-raising on the part of Government officials and the public at large as to the food security situation of many Brazilians and the numerous measures at hand to seek redress.

10. Capacity Building

Experience has shown Brazilians that any progress towards realizing the right to food requires a team of informed, capable and willing participants, all pulling in the same direction, with shared goals. It has also shown that when activities produce the desired outcome while simultaneously stimulating public interaction, a social process takes place that is both self-reinforcing and sustainable.

The drafting of LOSAN in 2006 demonstrated just this kind of teamwork. While the one-year effort resulted in a better law, incorporating strong right to food language, it also meant that members of government, CONSEA and civil society participated in an intensive right to food learning experience while they worked.

Brazil's MDS commissioned civil society and human rights experts to create a right to food distance learning course.[60] More than 2 000 'learners' completed the first edition of the course.

Its targets are government officials, civil society organizations, members of CONSEA and social control councils. A second edition was launched in 2009.

Training courses in right to food were offered to counsellors of state CONSEAs, public executives, civil servants and community residents. This whole discussion on the right to food contributed to the start of a capacity development process, which is fundamental for the realization of human rights. In Piauí, for example, the state CONSEA board expressed interest in intensifying the debate on the right to food and this contributed to the actions developed by the Council. Rights defence institutions, such as the Federal and State Public Prosecutors of Piauí and Alagoas, intensified the use of judicial and administrative instruments to protect the right to food. These workshops encouraged states to replicate CONSEA's Standing Commission on the Human Right to Adequate Food so as to make existing recourse and claim mechanisms available at state level. General right to food knowledge benefits all stakeholders; however, officials in charge of implementing policies and programmes need specific training.

CONSEA and ABRANDH staged a national Right to Food Guidelines Campaign which included translating, publishing and distributing the Right to Food Guidelines (more than 20 000 copies) along with an explanation booklet. The alliance of Brazil's civil society groups – the National Forum on Food and Nutrition Security (FBSAN http://www.fbsan.org.br) launched an information website that hosts a chat room for exchanging ideas and experiences. Alliances between NGOs,

---

60   http://www.direitohumanoalimentacao.org.br

journalists and Brazil's *Radiobrás* news service helped disseminate facts about the rights and responsibilities of individuals, and the obligations of the government.

Development projects in the two slum areas of Sururu de Capote and Vila Santo Afonso, described above, held capacity building workshops that increased understanding of human rights and the State obligations involved, and informed participants about right to food claim and recourse methods. As a result of becoming knowledgeable about their children's food insecurity and their right to make demands on public officials, residents were enabled to lobby for changes in health care, education, basic sanitation and food assistance – for instance, during a public hearing facilitated by the Public Prosecutor.

Dialogue between inhabitants of Vila Santo Alfonso and public officials of the municipality of Teresina (state of Piauí) was particularly tense during a public meeting in July 2005. Local government officials argued that they should not be expected to deliver remedies as only the President could deal with the country's poverty. With an assertiveness acquired through a year of advocacy, one community leader stood up and told them: "Some days I have only one meal and sometimes only rice. You are our Government representatives so you are responsible. You have to provide this to us because it is our right." In spite of the fact that the inhabitants of this slum still live in difficult conditions, the people feel that they are no longer invisible.

As the Government of Brazil increases the capacity of the agencies involved in planning and implementing human rights based public policies, the experience of FAO, other UN agencies, and other international organizations, can be of assistance. International organizations, in turn, can learn from Brazil's experiences and help share this knowledge where it can benefit others.

Awareness raising and capacity building paid off when the third National Food and Nutrition Security Conference in July 2007 discussed the design of Brazil's new Food and Nutritional Security System. From Cabinet insistence that social programmes constitute rights rather than charity, to counsellor requests for better claim mechanisms; and from tribes in remote villages, to the thousands crowding the convention centre to take part, awareness of the right to food was undeniable.

11. Conclusions

In looking at the Brazilian government's efforts to promote, protect and ultimately deliver on its promise to realize the right to food, it seems obvious that political commitment by the executive branch of government is key. However, there is commitment to create capacity for Brazilians to participate in the *process* of the realization of the right to food and to do so at community level – not just at state or national levels. To have a Presidential commitment to end hunger and protect the right to food, was surely a driver, but the success of Brazil lies on its support of civil society activity – legal, financial and political. This included reaching out to the poorest of the poor, identifying their needs and concerns through disaggregated data collection, and mapping of vulnerability. Concretely, this meant reaching out to remote rural areas where people are 'invisible' as far as political power is concerned as well as granting citizens eligibility to benefit from social programs.

Along with government efforts to promote social inclusion and involve the citizenry in the fight against hunger, one can also see a true willingness of informed citizens to mobilize and hold government officials accountable for the right to food. This is seen in many spheres of activity – be it designing social assistance programs, delivering school meals, identifying who is eligible for family grants, monitoring violations of rights, and providing remedies for the poor and hungry. Part of this is due to systematic knowledge dissemination and capacity building through different tools such as handbooks on the right to food for government officials or chat rooms for dialogue on rights. However, such mobilization is facilitated by the authority given to institutions like CONSEA, which plays a key role in directing policy design and implementation, and which is composed by a 2/3rd majority of civil society members.

Nevertheless, in Brazil, the participation of civil society is not limited to public policy design. It can be seen also in social control mechanisms throughout the implementation process. Civil society is often also involved in the demand for recourse in cases where implementation is hampered by violations. The National Rapporteur on the Right to Food, Water and Rural Land in Brazil serves as a key player in putting right to food on the national agenda. He does that through his authority to investigate violation claims. In teaming up with legal officials, such as the Public Prosecutors and the Special Secretariat for Human Rights, the Rapporteur becomes a political tool for civil society to demand accountability and transparency from the local, state and central government. This would normally create a level of tension on the national political scene; however, one can not underestimate the level of cooperation that exists in Brazil, a sort of partnership so to speak, between government and civil society. Nor can one ignore the level of coordination between municipal, state and federal government bodies to create an environment in which right to food is a central guiding objective – and thus responsibility – in the overall food security strategy adopted by Brazil – including LOSAN – and in the permanent food security governance structure of the country. Therefore Brazil is a leading example of how the right to food can overcome the obstacles in social, political and economic development that would otherwise sabotage efforts to end hunger.

Recommendations

Civil society and the Government of Brazil are pioneers in promoting the implementation of the human right to adequate food. As this case study shows, Brazil has experienced enormous progress in all seven steps of the implementation process. The recommendations from this case study would be:

- To promote civil society organization, mobilization and participation in order to guarantee that: (i) the plight of the most affected is visible; (ii) the neediest are empowered; (iii) public action is transparent; (iv) goals and priorities are achieved and monitored; and (v) accountability mechanisms are promoted.
- To promote human rights approach as fundamental to (i) diagnosing the food security situation and identifying the most vulnerable; (ii) facilitating the establishment of priorities; (iii) empowering the most affected right holders; (iv) ensuring broad and meaningful participation of the most vulnerable; (v) clarifying State obligations; and (vi) guaranteeing the right to make a complaint.
- To support the establishment and financing of independent recourse instruments, such as the Public Prosecutor, public defenders and national human rights institutions, the National Rapporteur on the Right to Food, Water and Rural Land as key players in the implementation of the right to adequate food.

III. GUATEMALA
Writing a Page of History

Main Points

- Guatemala was one of the first Latin American countries to have a right to food law. In recent years many more countries have followed its example.
- The institution of a Presidential Commission on Human Rights has been a valuable asset in fostering the realization of human rights.
- Right holders need to be adequately informed about human rights and their right to claim in cases of non-fulfilment. Equally important is the development of capacities of duty bearers, to ensure that they are familiar with their respective obligations and act in a responsible manner.

1. Background

Guatemala has sufficient legal tools to guarantee the enjoyment of the right to food for all its people. Article 46 of the Guatemalan Constitution[61] states that in the matter of human rights, international law prevails over domestic law. The Congress of the Republic of Guatemala ratified the ICESCR in 1987 (Agreement No. 69-87). Article 1 of the ICESCR states that *'In no case may a people be deprived of its own means of subsistence'*. In 2005, Guatemala promulgated Legislative Decree 32-2005 establishing the National Food and Nutrition Security System (*Sistema Nacional de Seguridad Alimentaria y Nutricional* – SINASAN). The enactment of this law and the adoption of a National Food Security Policy in 2006 are a major step forward, opening up greater possibilities for the government to respect, protect and fulfill the right to food.

Hunger and malnutrition, which are so widespread in Guatemala, cannot be explained simply in terms of lack of food. In theory, the country has sufficient land to feed the whole population without difficulty. The problem is due to an unequal distribution of the country's productive resources. Land and wealth are concentrated in the hands of a few, due to a long history of development practice that excluded the majority of the people and has left an enormous number of small farmers, most of them indigenous people, either landless or without labour rights.[62]

An analysis of the social and economic impact of infant malnutrition in Guatemala published by World Food Programme (WFP) and the Economic Commission for Latin America and the Caribbean (ECLAC) in 2005,[63] quantified the cost of malnutrition in terms of its negative impact on health, education and productivity.

---

61 Constitution of the Republic of Guatemala, Article 46, Primacy of International Law. This article enshrines the general principle that in the matter of human rights, the treaties and conventions accepted and ratified by Guatemala take precedence over domestic law.

62 Commission on Human Rights, 62nd Session. 2006. The Right to Food. *Report of the Special Rapporteur on the Right to Food, Jean Ziegler. Mission to Guatemala.* UN doc. E/CN.4/2006/44/Add.1.

63 WFP and CEPAL. 2005. *Hambre y desnutrición en los países miembros de la Asociación de Estados del Caribe.* Serie Políticas Sociales 111.

The survey estimated that the total cost of malnutrition in 2004 was 24 853 million quetzales or US$3 128 million, accounting for 11.4 percent of GDP and 1.85 times the country's social expenditure that year.

In 2004, the national health system assisted 560 000 patients, including those suffering from malnutrition, marasmus and kwashiorkor. Illnesses caused by diarrhea-related infections, acute respiratory infections and anemia cost the State US$285 million, accounting for about 9 percent of the total cost indicated above, and 1.17 times their public health expenditure.

2. Identifying the Hungry in Guatemala

The task of monitoring and evaluating the state of food security and nutrition in Guatemala is the responsibility of the Food and Nutrition Security Information and Coordination Centre (*Centro de Información y Coordinación de Seguridad Alimentaria y Nutricional* – CICSAN). The main duties of this Centre are to monitor the state of food security and nutrition (including availability, access, consumption and utilization of food); to ensure cooperation among institutions working in the field of food security and nutrition throughout the country, and to create an early warning system to monitor the risk of food insecurity. It also facilitates links between the authorities and communities facing problematic situations, through a rapid response mechanism.

CICSAN has demonstrated its reliability and, in some cases, has become a regional benchmark for Central America. The Executive 2004-2008 ensured that in 2006 approximately US$2.5 million were allocated specifically for restructuring the Centre. CICSAN was very successful in collecting information on the vulnerability of each community, through the Municipal Information System on the Risk of Food and Nutrition Insecurity.

The country's efforts to take stock of the right to food situation also involved the Office of the High Commissioner for Human Rights (OHCHR), several social stakeholders[64], the Human Rights Ombudsman, the Presidential Commission on Human Rights, the Ministry of Public Health and Social Welfare[65] and FAO Guatemala, through the members of the Special Programme for Food Security and the Right to Food Project in Guatemala.

When a human rights based approach is used to identify the high prevalence of chronic infantile malnutrition, all programmes, policies and technical assistance provided by the UN System should promote the realization of the right to adequate food, since human rights standards and principles should serve as a guide in every sector and in all phases of the planning process involving international cooperation.

There is a growing need for other players with social responsibilities to take part in seeking solutions to make the right to food effective. The business community in particular should take part in this process. In the long term, entrepreneurs will also be affected if the situation in Guatemala continues to deteriorate. The corporate sector influences the framing of policies

---

64   One of the most important of these civil society bodies is the International Centre for Human Rights Research which not only takes part in designing the indicators but has issued the largest number of publications on the Right to Food of the Guatemalan people.

65   Key government-level parties were invited to take part in drawing up the methodology and defining the criteria, but only the Ministry of Public Health and Social Welfare responded to the invitation.

throughout the whole of Latin America and nowhere more than in Guatemala. The influence of the sector in these countries indicates their ability to alter the course of important economic policies, such as the taxation regime.[66]

3. Sound Food Security Policy

Although there is a sound understanding of the concept of food security and nutrition in Guatemala, the human rights component is not widely understood. Food security is considered an elementary concept and has more meaning for the people than information on their human right to adequate food.

When drafting a convincing food security strategy people do not always consider human rights as a key element, rather, drafters generally refer to a reduction in chronic malnutrition, based on FAO's traditional 'four pillars of food security' (availability, stability of supply, access and utilization).

The analysis of international experiences showed that the programmes designed to reduce malnutrition are routinely based on a model of nutrition security that envisaged health improvements in a healthy environment and with good health care and nutrition practices in the home. Programmes with a food security focus, on the other hand, are based on a model giving pride of place to food availability and access. This is a reflection of the way different sectors are set off against each other both by governmental institutions and by cooperation agencies, with the result that in some cases they find it difficult, in their conceptual and operational patterns, to combine both issues. This situation demands effort to ensure that the work is carried out in a coordinated manner.[67]

One example of coordinated efforts within governemntal institutions was set in motion in May 2005 when the Presidency's Secretariat for Food and Nutrition Security (*Secretaría de Seguridad Alimentaria y Nutricional* – SESAN) in Guatemala set in motion a process of inter-institutional consultation and planning. This led to the drafting of the Programme for the Reduction of Chronic Malnutrition 2006-2016, based on an essentially pre-emptive strategy. The implementation of the programme is a concrete demonstration of the government's intention to progressively realize the right to food. As Andrés Botran, Secretary of SESAN (2005-2007), stated: *'The overall national goal is to halve chronic malnutrition by 2016. The main objective is to reduce chronic infant malnutrition by 24 percent nationwide during the first nine years. This programme is based on the principles of focus, prevention, comprehensiveness and sustainability.'*

Programme implementation began in 2006 with the setting up of a pilot scheme in 18 municipalities. The scheme was expanded during the same year to cover 83 municipalities, with the following components: basic health care services, food and nutrition education, breast-feeding and supplementary feeding. By 2007 it was planned to introduce the feasibility and sustainability components (basic water and sanitation, community organization, improving the household economy) in 17 of the 83 municipalities.

---

66   Inter-American Development Bank. 2006. *La política de las políticas públicas*, Chapter 8.

67   *Evaluación, actualización y fortalecimiento del programa para la reducción de la desnutrición crónica 2006/2016* by Andrés Botran with the support of his technical team, financed by FAO, UNICEF and USAID. This report was made public in December 2007 and presented to the Food and Nutrition Security Secretariat of the Presidency of the Republic as an inter-institutional contribution, proposing that the efforts continue.

Although this strategy does not embrace all the right to food principles, it is a means of implementing activities that will greatly reduce the number of hungry people in the country. As people become more aware of their right to food, the strategy should be implemented with the participation of all sections of society.

4. Allocating Roles and Responsibilities

The SINASAN Law incorporates three tiers of action: **policy-making**, represented by the National Food Security and Nutrition Council (CONASAN); **technical planning and coordination**, which falls within the remit of SESAN; and **implementation**, which comes under the ministries and organizations with operational responsibilities. SINASAN is composed of CONASAN and SESAN, as well as the Consultation and Social Participation Authority (INCOPAS) comprising representatives of 10 social sectors, and the Support Institutions Group (GIA) that provide consultancy services to the Secretariat.

The organizational chart below indicates the different bodies involved in Guatemala's National Food and Nutrition Security System.

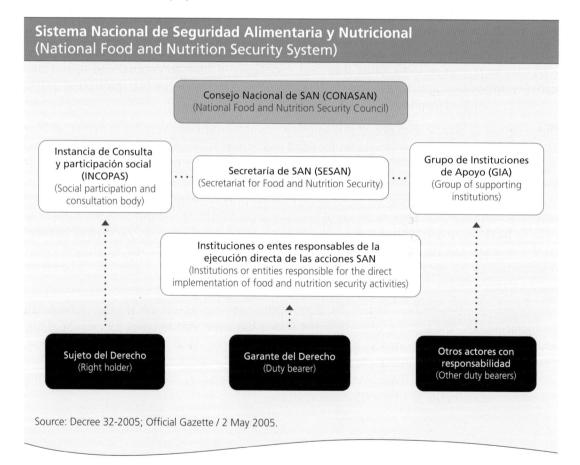

Sistema Nacional de Seguridad Alimentaria y Nutricional
(National Food and Nutrition Security System)

Consejo Nacional de SAN (CONASAN)
(National Food and Nutrition Security Council)

Instancia de Consulta
y participación social
(INCOPAS)
(Social participation and
consultation body)

Secretaría de SAN (SESAN)
(Secretariat for Food and Nutrition Security)

Grupo de Instituciones
de Apoyo (GIA)
(Group of supporting
institutions)

Instituciones o entes responsables de la
ejecución directa de las acciones SAN
(Institutions or entities responsible for the direct
implementation of food and nutrition security activities)

Sujeto del Derecho
(Right holder)

Garante del Derecho
(Duty bearer)

Otros actores con
responsabilidad
(Other duty bearers)

Source: Decree 32-2005; Official Gazette / 2 May 2005.

CONASAN is the governing body of SINASAN and is responsible for promoting food and nutrition security in the national political, economic, cultural, operational and financial sectors.[68] The decisions taken and agreements reached at the Council meetings are binding on its member institutions, through their official representatives. The representatives of member institutions are also responsible for ensuring that the institutions they represent comply with the instruments and perform the actions prescribed in the Food and Nutrition Security Policy: that they follow-up on the activities stemming from the strategic and operational plans, in order to address serious contingent food insecurity problems; and that they comply with other instructions and guidelines issued by resolution. These have to be implemented by governmental institutions as soon as they are enacted by SESAN[69] in its capacity as Council Secretariat.

One very important responsibility vested in CONASAN by the SINASAN Law is to recognize, analyze and propose adjustments to food and nutrition security policies and strategies, based on the annual recommendations issued by the Human Rights Ombudsman to honour, safeguard and progressively realize the right to food and nutrition security.[70]

SESAN is SINASAN's coordinating body, responsible for the interdepartmental, operational coordination of the Food Security and Nutrition Strategic Plan, and for linking the programmes and projects of the national and international institutions involved in national food and nutrition security.[71]

It is also responsible for setting out the procedures for technical planning and coordination between State institutions, Guatemalan society, non-governmental organizations and international cooperation agencies involved in food and nutrition security at national, departmental, municipal and community levels. One of SESAN's key responsibilities has been the design and operation of SINASAN – to monitor and evaluate the state of food and nutrition security, the effects of strategic plans and programmes and also the early warning system, so as to identify contingent food and nutrition insecurity situations.

The SINASAN Law is structured in terms of roles and functions. In the case of the duty bearer, it states that ensuring food availability is the responsibility of the Ministry of Agriculture, Livestock and Food, in coordination with other institutions; the aim is to encourage activities that will contribute to making food that is both suitable and safe – either locally-produced or imported – permanently available to the people.

The tasks of guaranteeing access to food and the stability of food supplies are the responsibility of the Ministry of Agriculture, Livestock and Food, the Ministry of the Economy, the Ministry of Work and Social Security and the Ministry of Communications, Infrastructure and Housing. This undertaking is carried out through encouraging activities that contribute to the people's physical, economic and social access to food on a stable basis.

---

68 Legislative Decree 32-2005. *The SINASAN Law, Article 12, on the nature of CONASAN.* (Available at http://www.sesan.gob.gt).

69 *Ibidem, Article 14, on the responsibilities of the members of CONASAN* (Available at http://www.sesan.gob.gt).

70 *Ibidem, Article 15 (j), on the powers and responsibilities of CONASAN.*

71 *Ibidem, Article 20, on the nature of SESAN* (Available at http://www.sesan.gob.gt).

Food utilization is the responsibility of the Ministry of Public Health and Social Welfare, the Ministry of Education, and the Ministry of the Economy, in coordination with other public agencies. These ministries are responsible for developing people's capacity to take appropriate decisions regarding food selection, conservation, preparation and consumption; encouraging activities that will enable people to maintain an adequate level of environmental health and hygiene in order to make best use of the nutrients contained in the food they consume; and implementing activities to strengthen and continually update human and institutional resources, including personnel from other agencies, regarding diagnosis and consequent treatment, recovery and rehabilitation of people suffering from malnourishment.

Similarly, urban and rural development councils need to establish specific food and nutrition security commissions in their departments, municipalities and communities. These commissions should act in compliance with the objectives of the Food and Nutrition Security Policy and the Strategic Plan, in coordination with SESAN.

5. Legal Framework to Realize the Right to Food

The development of a legal framework for the realization of the right to food has been a long process. It started in 1975 with the Guidelines for a National Nutrition Policy (*Lineamientos Para Una Política Nacional de Nutrición*). It was not until the end of the 1990s that more frequent attempts were made to institutionalize Food and Nutrition Security (FNS). This gap in the series of initiatives relating to people's food security broadly coincided with the years of domestic armed conflict, which ended with the signing of the peace agreements in 1996.

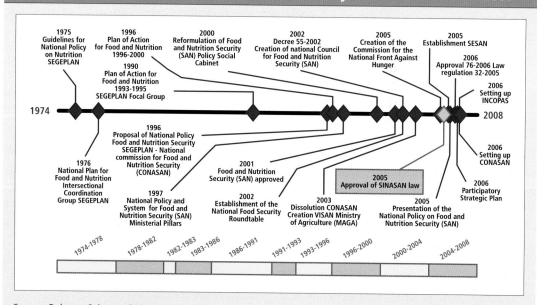

Institutionalization of Food and Nutrition Security in Guatemala 1974-2007

Source: Roberto Cabrera, GIISAN, Guatemala

Compliance with the commitments made by signing national and international agreements and conventions[72] became more urgent when the 2001 coffee crisis and drought triggered a dramatic food emergency, placing the national rural population's means of subsistence in jeopardy. This crisis also revealed that the people were extremely prone to food insecurity;[73] consequently, it was an opportunity to be exploited for the purposes of FNS.

The above events provided the motivation for social groups, government and international cooperation to combine their efforts. This pooling led to the publication of Ministerial Agreement 577-2003 establishing the National Food Board, comprising social leaders and government representatives, and supported by international cooperation organizations. The Board members were to work together to produce the framework for a State food policy which would address the issues of food sovereignty, adequacy and nutrition. Inputs would be provided by organizations and communities of indigenous people, civil society, producers and government, accompanied by the media, the Human Rights Ombudsman and international cooperation organizations. The National Food Board was co-chaired by the Ministry of Agriculture and the leader of an indigenous smallholders' group. It was intended to become a forum for dialogue, issuing decisions that were binding on the sectors linked to it.

The first attempt to dissolve the Board came when sections of the business community endeavoured to have the Ministerial Agreement declared unconstitutional, on the grounds that civil society is not responsible for making public policy. The Constitutional Court found in favour of the claimants. Despite losing its statutory status, the Board continued working with its few remaining resources. It reached a peak during the period 2003-2005,[74] when the National Front against Hunger (representing the Executive), and the Food Security Legislative Commission (representing Congress), together with the Inter-Institutional Food and Nutrition Security Group (GIISAN) representing a critical group of technicians and experts, supported it on a voluntary basis.

Because of the fragmentation of human and financial resources in Guatemala, governmental and international cooperation programmes and projects have not had the intended effect on the prevalence of chronic malnutrition and poverty.[75] Therefore, the original purpose behind the SINASAN Law was the inter-institutional coordination of interdepartmental and international cooperation actions concerning people's food and nutrition security. The law was approved in 2005. Initially, the implications of acknowledging the human right to adequate food in an ordinary Act of Parliament were neither clearly identified nor fully understood. Only a small number of specialists were aware of what was at stake. This is another reason why no amendments were

---

72 (i) The signing of the Peace Accords. 1996. Guatemala; (ii) The International Conference on Nutrition. 1992. Rome; (iii) The World Food Summit. 1996. Rome; (iv) and CESCR, *General Comment 12*. 1999. Geneva.

73 UNDP. 2002. *Food Security and Nutrition Status Survey*. Guatemala.

74 2003 was an election year in Guatemala, and the Board succeeded in having the two candidates for the Vice Presidency of the Republic, from the strongest political parties, sign a letter committing them not to remove the issue of food security of the Guatemalan people from the political agenda. This letter was used as support when Oscar Berger took over the Presidency in 2004.

75 In Guatemala, out of a total of 6.6 million poor people (51 percent of the Guatemalan population have an annual income of US$876) 1.9 million are extremely poor (15.2 percent of the total population earning US$427) according to the *Living Standards Survey* (ENCOVI 2006).

made to the key articles of the Act during its passage through Congress. The perspective has since changed and today's aim is to realize human rights.

Many beneficial results have stemmed from the adoption of the SINASAN Law.[76] Some of the actions fall under the Food and Nutrition Security Secretariat of the Presidency of the Republic, but all are performed within the framework of SINASAN. Among the most important of these are:

- The integration of 81 Municipal Food and Nutrition Security Commissions and 17 Departmental Commissions, the latter benefiting from institutional strengthening.[77] These Commissions operate internally, on the urban and rural development councils, based on the Devolution Act. Their main task is to help in linking the activities of the operational branches of ministries and local NGOs.[78]
- The establishment of the Food and Nutrition Security Information and Coordination Centre (CICSAN). The Centre was established to provide a common forum to integrate, process, analyze and disseminate useful information on food and nutrition security as a tool for decision-making, fostering agreements and coordinating focused activities. It draws on different national public information sources and on international cooperation agencies.
- Following Decree 32-2001[79], the specific budgetary allocation[80] of one-half of one percentage point (0.5 percent) of VAT came into force, specifically for food security programmes destinated for poor and vulnerable persons. The Decree explicitly states that this percentage figure should not be construed as being the budgetary 'ceiling' but rather the budgetary minimum. The Ministry of Finance, through the Technical Directorate of the Budget (DTP), must include this allocation in the central government's General Budget every fiscal year.[81]

The SINASAN Law provides an opportunity for applying a human rights based approach. Based on non-discrimination, transparency and participation, the law acknowledges certain principles which apply to food security policy, such as food sovereignty, ownership, sustainability, prudence and decentralization. All this is done in the light of the right to adequate food, which is mentioned frequently in the provisions of this law. These provisions mark a genuine step forward, although they leave an open question in relation to right holders and duty bearers: Who is responsible for training duty bearers to implement public policies and plans, and to be aware of the implications and the full scope of the law?

The Delegation of the European Commission created an effective method for generating awareness of the SINASAN Law in the media, through its projects in Guatemala. The European

---

76   Vera Scholz Hoss & José Luis Vivero Pol. 2008. *Are Food Security Acts Useful for Reducing Malnutrition?* Analysis of the Guatemalan Case, ALCSH Working Paper. Santiago, Chile.

77   Government of the Republic of Guatemala. 2007. *Report on the State of Compliance with the Commitments Undertaken at the 9th Regional Conference on Latin American and Caribbean Women*. Quito, Ecuador.

78   Vivero, J.L. y Monterroso, I.E., 2008. *Comer es un Derecho en América Latina. Avances legales y políticas a favour del derecho a la alimentación*. ALCSH Working paper # 03. Santiago, Chile.
      (See http://www.rlc.fao.org/iniciativa/pdf/wp3.pdf).

79   This decree reformed the Value-Added Tax (VAT).

80   This is one of the most important points with reference to SINASAN because it explicitly refers to a minimum value for the annual allocation of resources.

81   See Article 38 of SINASAN referring to the specific budgetary allocation.

Commission trained approximately 150 communicators – editors, announcers and journalists – to circulate the relevant messages through the municipal community radio in the Department of Huehuetenango. Graduates of the training course received a diploma. After the first year, a survey showed that 41 percent of the people interviewed knew that they had the right to nutritious and healthy food. They had been given this information by ACODIHUE,[82] SESAN and the *Creciendo Bien* [83] programme. Forty-eight percent said the broadcasts helped them to improve the quality of their food four times a week on average; 86 percent understood the message; and 4 out of 5 who heard the message put it into practice.

Despite the success of the SINASAN Law, there are still a number of challenges to be addressed:

- defining the role of the governing body of CONASAN, chaired by the Vice President of the Republic, because provision was made for putting political mechanisms in place to facilitate inter-sectoral coordination;
- joint planning by all the institutions forming part of SINASAN, as required by the law;
- strengthening the technical and administrative structure of SESAN, increasing its operating budget;
- providing greater political backing to SESAN's departmental delegations and its decentralized inter-sectoral communications;
- increasing the impact of INCOPAS on SINASAN and advocating the establishment of monitoring and social auditing mechanisms;
- implementing the Food and Nutrition Security Policy in a harmonious and coherent manner; and
- formally establishing the GIA and setting a common international cooperation agenda.

6. Monitoring the Right to Food

Civil society in Guatemala plays a vital role in ensuring that the institutionalization of the right to food is sustainable in the long term. Civil society organizations now openly state that a human rights based approach is their priority.

The duty bearer must fulfil the obligations to respect, protect and fulfil but the right holder must also demand compliance. There has been considerable progress in this area in different sections of society, and there are several reports of compliance with ICESCR.

The report by the Special Rapporteur on the Right to Food, following his visit to Guatemala in 2005, is now being followed up. Articles are being published in the press, and information disseminated to rural communities with regard to possible violations of fundamental rights, while increasing progress is being made with human rights based monitoring and the implementation of the Right to Food Guidelines. As a result of the technical and documentary work undertaken, a political impact is now being felt in the different areas with regard to the realization of the right to food. To mention but one example, a revealing investigation was conducted by *Fondo de Tierras y del Vaso de Leche*, using a human right to food based approach, recording the

---

82   Asociación de Cooperación al Desarrollo Integral de Huehuetenango is a European Commission project in Guatemala.

83   First Lady's Social Works Secretariat Programme.

perception of the local participants in the different programmes. Parallel reports are drafted each year, and the recommendations by the Special Rapporteur are being monitored.[84]

Another monitoring mechanism provided by the SINASAN Law is the annual report submitted by the Ombudsman to the National Food and Nutrition Security Council. This report became a statutory obligation by decree 32-2005[85] and provides an opportunity to follow up on these policies. A fact-finding report was drawn up[86] so that adjustments can be made in relation to Guatemala's compliance with the obligation to respect, protect and fulfill the human right to adequate food.

The Ombudsman is gradually taking a leading role and has reported cases of mismanaged programmes and mismanagement in areas such as fertilizers in exchange for political favours. He has also been monitoring the death certificates of individuals whose deaths were linked to malnutrition. This work is welcomed by the mass media. They give particular prominence to news regarding policy adjustments and monitor compliance by government programmes or the targeting of activities. At present, more than 120 officials from the office of the Ombudsman have been trained by FAO's PROCADA project.

The Action Plan of the Presidential Commission on Human Rights (COPREDEH) indicates the right to food as a priority in its strategic plan to 2017. COPREDEH is a government entity that was created to improve coordination of activities between the Ministries of State, the Judiciary and the Office of Human Rights. Its main tasks relate to the centralization and monitoring of information on complaints of human rights violations and in promoting research through the Ministry of the Interior and the Public Ministry.

7. Legal and Administrative Recourse Mechanisms

The progressive realization[87] of the right to food and the implementation of justiciability[88] may be mutually supportive but there is a great difference in the time required to make them effective. To sue for recognition of a human right before a court in Guatemala takes an enormous length of time within the judicial system and the human rights defense mechanisms. Consequently, there have been almost no cases brought before the courts: not one of the 35 000 complaints

---

84   Centro Internacional para las Investigaciones en Derechos Humanos. 2007. *La Alimentación Un Derecho Desnutrido.* p. 47 (Available at http://www.ciidh.org). *Red Nacional para la Defensa de la Seguridad y Soberanía Alimentaria en Guatemala, Centro Internacional para las Investigaciones en Derechos Humanaos, Coordinación de ONGs y Cooperativas, y Pastoral de la Tierra Interdiocesana Provincia Eclesiástica de los Altos. 2007. Informe Alternativo del Derecho a la Alimentación en Guatemala – Monitoreo de las Directrices Voluntarias.*

85   SINASAN Law, Article 15, (j): *'To take cognisance of, analyse and propose Food and Nutrition Security policies and strategies, based on the recommendations issued each year by the Human Rights Ombudsman on complying with, protecting and progressively realising the right to food and nutrition security.'*

86   This fact-finding exercise is designed to ensure that the indicators are followed up yearly. Under the guidance of Edgar Ortiz, who elaborated the process, the exercise was carried out following a participatory process in which duty bearers, right holders and observers took part in drafting the first report that was delivered to the Vice President of the Republic and the Secretary of SESAN in August 2008.

87   *'The concept of progressive realization can be described essentially as the State's obligation to parties to a) adopt all relevant measures for the application or full realization of economic, social and cultural rights; and b) to do so allocating the maximum resources available.'* Report of the United Nations High Commissioner for Human Rights, Substantive Session, 2007.

88   FAO. 2006. *Guidelines on the Right to Food: Information Papers and Case Studies*, p. 79. Rome.

filed with the Ombudsman in 2006 referred to the right to food,[89] and it is believed that the Ombudsman's office examined at least four cases in 2008.

There is only one documented case of justiciability of the right to food in Guatemala. The judge in the Labour Court of First Instance in the Department of Quetzaltenango, Clara Diria Ezquivel, ruled in the final judgment of a labour dispute that a cleaner's right to food had been violated, and imposed a penalty for the violation. This was not the first instance of justiciability of the right to food in the country but the court revealed the fragile structure of the authorities responsible for ensuring the realization of this human right.

Excerpt from an article published in the local daily newspaper El Quetzalteco in June 2006

The papers in Case No.154-04-128 before the Quetzaltenango Labour Court brought by Carmen Janeth Molina and heard on 5 April 2004, contain news that will encourage all defenders of the right to food because this was the first case in Guatemala in which the court settled the case, finding that the defendant had violated the plaintiff's right to food. According to Carmen's account, the origin of this dispute was the refusal by her employer to pay her for her work. Carmen was responsible for cleaning and working in the reception of 'the company' when she was laid off. This was a typical case of the failure of her employer to fulfil his duties towards her but it also entailed a flagrant violation of her right to food, because the company's defense team had tried to [induce her to] give up any hope of successfully suing and used ploys and cunning to prolong the final judgment, spinning out the trial for 24 months. During this time Carmen fell into arrears with her rent and was hungry on several occasions because her only means of subsistence was this work. Unlike the company, the only support Carmen had received was from the legal aid office that had appointed assistants who had not yet graduated, to defend her.

In the judgment issued by the Labour Court of First Instance, the judge applied Article 11 of the International Covenant on Economic, Social and Cultural Rights (ICESCR), in which the States Parties recognized the right to food for all people. The ICESCR was backed up by Article 46 of the Guatemalan Constitution and was enshrined in an ordinary Act of Parliament in Decree 32-2005, enacting the Food and Nutrition Security Law.

This opened the door to greater respect on the part of the State as a whole towards the human right to food. Other judges will be able to follow the same path and argue, with the Law in their hands, that the right to food is enshrined in the ICESCR and that an ordinary Act of Parliament recently enacted the Food Security and Nutrition Law.

---

89   Data provided by Karla Villagrán, Director of the Investigation Unit of the Office of the Right to Food Ombudsman, December 2006.

The justiciability of the right to food is the right to invoke a human right, recognized in both general and theoretical terms, before a court or recognized quasi-judicial body. The objectives of this action would be, primarily, to establish whether the human right has been violated or not in a particular case laid before the institution and, secondly, to decide on the appropriate measures to be adopted in the event of a violation.[90]

The right to invoke a human right. In Guatemala, international human rights instruments are not considered by the general population to be a formal source of law, despite the fact that Article 46 of the Guatemalan Constitution lays down the general principle that in the matter of human rights, the treaties and conventions that have been accepted and ratified by Guatemala have primacy over domestic law. Because of the widespread perception that human rights are rather weak, the general public feel that there is little they can do when such rights are violated, and that complaints almost inevitably end up by being just another case among the pile gathering dust in a government office or court.

This is also the reason why the above-mentioned case regarding the justiciability of the right to food was not the subject of a court case (it was settled out of court) and why the injured party, Carmen Janeth Molina, was not even aware of the reasons behind the judgement when she was interviewed at her home.

Recognition in both general and theoretical terms before a judicial or quasi-judicial body. In July 2005, a two-day training course was conducted jointly by FAO, FIAN and the Supreme Court of Justice, for Supreme Court judges, SESAN delegates, members of GIISAN and representatives of civil society. The course was held to illustrate the progress made with regard to recognizing the right to food, beginning with Guatemala's adoption of the SINASAN Law (Decree 32-2005). One of the two-day training sessions was entitled 'The Right to Food: a Challenge to Justice.'

Judge Clara Diria Ezquivel attended on that day. Addressing the theme 'The application of law to food [issues] in agrarian and labour disputes in Guatemala,' she expressed her intention to examine a number of cases, 'strengthened', as she put it in her paper, by the recent enactment of the SINASAN Law. Although this first approach to the judges of the labour courts led to recognition in very general terms, the matter had already entered the sphere of a judicial body.

Decision as to whether or not a human right had been violated, in the specific case submitted to the court. This consideration was quickly supported by Judge Ezquivel. But when the judgment was announced, the defendant objected and appealed to the Supreme Court claiming inconsistency and favouritism. The initial claim was considered valid and the penalty to be imposed on the violating company was approximately US$6 500.

---

90 FAO. 2006. *Guidelines on the Right to Food: Information Papers and Case Studies*. p. 79. Rome.

8. Capacity Development

With the coming into force of the SINASAN Law, it became necessary to strengthen the capacity of duty bearers, right holders and observers of human rights. Technical teams from both FAO and SESAN were set up to develop a specific right to food, or right to food security and nutrition project for Guatemala. Accordingly, on 3 August 2007, at the 5th American Congress on the Agrarian Act, PROCADA was officially launched.[91] This project was set up in response to the conclusions reached by a group of food security and nutrition experts at a participatory consultation. When asked "What are the causes that keep the morbidity, mortality and human underdevelopment indicators associated with failure to realize the right to food, at these precarious and deteriorating levels?" the experts replied that the right to food was unknown.[92] Many of the lessons learned set out in this document were gathered within the framework of PROCADA.

FAO's PROCADA project focused on capacity building. Presentations on the human right to adequate food were included in the first awareness raising plan for mayors and municipal corporations (first semester 2005). Additionally, two postgraduate courses on Food Security and Poverty have taken place within the framework of the Special Programme for Food Security. Thirty-five specialized staff and experts from several government sectors, academia, civil society organizations and international development agencies attended the first postgraduate course. The second course was attended by 33 professionals and experts – also from government sectors, civil society organizations, academia and international development agencies.

The training modules on the right to food were structured with the assistance of the Presidential Commission on Human Rights, and the structuring process was participatory. The Office of the Attorney General for Human Rights and the OHCHR also played an important role in the progress achieved.

Training for the module on the right to food began in April 2008. In this first phase, a training for trainers course was held to promote a better understanding of the basic principles of the right to food, food security and nutrition. This initial training session was aimed at various institutions, such as the Assembly for Consultation and Social Participation, which appears in the SINASAN Law, the Office of the Attorney General for Human Rights and COPREDEH staff members, as well as several staff from the UN System. This resulted in a strong alliance between right to food actors, which later developed and led to action. A large number of people at central and regional levels received training in the Office of the General Attorney for Human Rights.

As a result of PROCADA's capacity building efforts, some stakeholders started to incorporate the right to food approach in their work. Examples include the left-wing network of municipalities which has begun to manage municipal projects with a focus on right to food, and also the Episcopal Conference in Guatemala (through Monsignor Alvaro Rammazzini), which includes the right to food theme as one of the four priorities in its workplan. In addition, several NGOs have begun work to

---

91 A project which forms part of the Right to Food Project implemented by FAO is Right to Food Unit in Rome, funded by the German Ministry of Food, Agriculture and Consumer Protection.

92 The conclusion reached by a group of technicians forming part of the Inter-Institutional Food and Nutrition Security Information Group (GIISAN), following a series of attempts to discover the reasons for the limited degree to which the Right to Food is exercised in Guatemala.

promote monitoring of public policies, based on human rights. Republican Congress benches will start auditing actions taken by the executive, with a human rights focus. Furthermore, institutions such as COPREDEH are holding internal training sessions to strengthen their capacity for follow-up work on cases taken to court and on the drafting of reports.

In 2008, the first right to food course was introduced as part of the curriculum for the degree course in nutrition at the Chemistry and Pharmaceutical Sciences Department of San Carlos University in Guatemala. The course was optional and the lectures were structured along the lines of the right to food module. Feedback from students was very positive and has opened the doors for practical training and research results that can be replicated. All the students enrolled in this faculty will undertake some months of community service as part of their practical experience, all of which will be supervised at different local levels.

A morning with you/your voice. PROCADA decided to involve more people in the efforts to spread information on the right to adequate food. Considering Guatemala's remarkable resource of young people, the key question was: How can young people be involved in the fight against hunger in such a way as to facilitate the formation of youth networks on the right to food? *A morning with you/your voice* is a programme aimed at listening to the young people and sharing views with them. This is a four-pronged working methodology: it captures attention, sensitizes, provides information and commitment.

The first step was a concert of music for young people, where information on the right to food and on the country's food security and nutrition situation was provided. Two songs highlighting these problems were composed for the occasion by local bands. The positive response motivated subsequent initiatives such as 'community coexistence' where groups of students live for three days and two nights in households vulnerable to food insecurity. The students must not take any money with them, only enough for the return bus ride. There is one young person per community. He or she notes down the points that concern him/her most and then makes a presentation for the other students.

The groups of young people who had had this experience returned soon afterwards with projects for community development in the communities that had hosted them. Moreover, their knowledge and sensitivity on the subject had increased considerably. The students identified the government officials directly responsible for work at community level and worked together with them to sensitize and familiarize them with current legislation on the implementation of right to food and its implications.

The media is often used as a tool for mounting political pressure. Apart from its usefulness in awareness raising, the media should also be utilized to create political will and add pressure. It frequently happens that what does not appear in the media does not exist on the political agenda. National communication media reflects public opinion while, at the same time, influencing the priorities on the political agenda and thus generating social and governmental reaction. Consequently, the role of the media is crucial for the realization of the right to food.

In February 2008, the first 'training breakfast' was held for the media. Surprisingly enough, 17 journalists from different sectors participated. The topic receiving the most media attention was chronic under-nutrition indicators at the municipal level – an issue of particular relevance as there are so many children suffering from malnutrition. On that occasion, several radio stations broadcast a news flash on the subject every 30 minutes. Consequently, the right to food was mentioned more frequently in one day than had been the case in several years.

This initial process involved frequent reference to the right to food, during which PROCADA became one of the established sources for consultation with regard to current affairs, such as agricultural production, under-nutrition, food policy and other matters. Written reports, as well as radio, television and digital productions promoted the idea that non-compliance with a human right would have an impact on the actual figures of underdevelopment. This increased communication with the media lead to the coverage of other requests from civil society, just as individual investigations by different reporters have resulted in positive reactions from the government.

As a result of media activities, Congress discussed possible decisions to improve food reserves, and the utilization of land for internal production and self-sufficiency. Most importantly, these activities provided momentum for central government to react to the food crisis.

9. Conclusions

It is both possible and feasible, financially and methodologically, to realize the human right to adequate food in Guatemala. It simply requires combining the goals of all stakeholders, tweaking plans, budgets and programmes, and working with international cooperation organizations, pursuing knowledge, encouraging attitudes towards change, and implementing processes that demand difficult adjustments.

International pacts, covenants and agreements, domestic legislation, methodologies, strategies, and recourse mechanisms are powerful tools but they require an 'engine' to drive them ahead.

Every country should have at least one 'engine of change'. That 'engine' must be a leader who is convinced that it is possible to realize the human right to adequate food. This person can be found among those holding public office, among the populations affected, or among organizations that take a leading role in monitoring and promoting the right to adequate food.

The will is lacking when the goal is not known: individual leaders – working on behalf of others or spurring them into action – must act as drivers and engines and join in efforts with other players to pursue a clear-cut goal. Above all, they must believe it possible for everyone to be able to afford to buy an adequate plot of land on which to produce food today, tomorrow, or in the future. Likewise, in emergencies, everyone should receive wholesome and varied food donations, not just any sort of food to fill empty stomachs, but nutritious food that is in keeping with people's traditional cultures and preferably produced in their own country. Steps must also be taken to ensure that food deliveries are not hampered by impassable or non-existent roads.

Everyone, at all times, must have the wherewithal to be able to feed themselves – this is the ultimate objective of the journey. What should be decided first and foremost is the intended objective and the final goal; methodology, accountability mechanisms, legal frameworks, plans and programmes can be fine-tuned later.

The enactment of the SINASAN Law may trigger activities and become the epicentre of structural change, or it may simply turn out to play a role fraught with good intentions. A great deal of hard work was put into achieving the enactment of this law in Guatemala and many politicians hailed its promulgation as a 'mission accomplished'. However, it has not yet fulfilled its mission, and its enactment was merely the first of many steps that still have to be taken.

One lesson has been learned, however: the first few years spent working on the progressive realization of the human right to adequate food in Guatemala has shown that it is essential to have a goal, the right attitude, and a driver. **The goal**: to ensure that all the people in Guatemala can feed themselves adequately at all times. **The attitude**: to be convinced that this is possible. **The driver**: a person (or group of persons) who steadfastly believes in the possibility of attaining this goal, makes it his/her raison d'être, and convincingly engages others to commit themselves to its achievement.

Recommendations

- Public spending on food security and nutrition programmes alone does not guarantee improvement of right to food situation. It is imperative that programmes apply human rights principles, such as participation and empowerment in the development and implementation processes to guarantee that right holders are empowered and enabled to claim their rights.
- It is crucial to keep the right to food in the public eye to guarantee that it does not disappear from the political agenda, and to build the political will through demands of right holders. This is so because the intensity of demands increases likelihood of decision makers to pay attention to them.
- Key data indicators on malnutrition and policy developments are generally discussed in the Ministries, institutes and cooperation agencies in central locations, such as the capital of a country. Such information must also flow down to mayors and officials at municipal and local levels to address the knowledge gaps of those living and working alongside the hungry people, and whose capacity needs to be strengthened so that they can fulfil their obligation as duty bearers.
- To address the constraints of lack of jurisprudence, public hearings should be encouraged to initiate legal processes. Also, closer monitoring of the legal process once initiated is required in such instances.
- The adoption of a law recognizing the right to food, establishing the institutions that will lead to its implementation, providing such institutions with a mandate and budget may be key. However, the law cannot be relied upon as the solution by itself. Political will and social organization and mobilization are required for the full realization of the right to food. These elements must be recognized and supported as the drivers of success in the full realization of the right to food.

IV. INDIA
Legal Campaigns for the Right to Food

Main Points

- The right to food is justiciable at the national level.
- Public interest litigation improves poor people's access to justice.
- Courts establish monitoring mechanisms to follow up on petitions.
- Legal entitlement to guaranteed employment improves enjoyment of the right to food.
- Food self-sufficiency has not translated into enjoyment of the right to food for all.
- An active civil society is necessary to strengthen implementation of programmes and court orders.

1. Background

India is a country where the State has intervened to ensure food security ever since independence. It has achieved remarkable production growth in the last few decades, and enormous economic growth in the last decade in particular. Yet, malnutrition rates are among the highest in the world, especially for children. The rich 'institutional landscape' of State and civil society actors combined with overall food sufficiency at the national level is in contrast with the very high figures of malnutrition all over the country – the Indian paradox, sometimes described as 'scarcity amidst plenty'.

India provides an example of justiciability of the right to food at the national level. The Supreme Court in Delhi has been addressing a public interest litigation case on the right to food since 2001 and has issued numerous interim orders creating legal entitlements to food and work under various Government programmes. The unique combination of strong public campaigning and direct and explicit orders from the Supreme Court has led to strengthened delivery of public assistance schemes and has greatly increased the accountability of public officials. This campaign also led to new legislation guaranteeing rural employment for the poor, which has been hailed as a landmark.

A country of vast proportions, India occupies most of a sub-continent and has a surface area of 3,287 thousand square km. It is the second most densely inhabited country in the world, with a population of more than 1.1 billion.

The Human Development Index for India in 2005 was 0.619, showing an increase from 0.551 in 1995 to 0.578 in 2000 (0=no development; 1=full development). In 2005 it ranked 128th out of 177 countries that had sufficient available information. In terms of equality, India is middle of the road with a Gini coefficient of 36.8 (absolute equality=0; absolute inequality=100).[93]

In the past, the country was frequently hit by large-scale, severe food crises, but there has been no major famine since independence. The Indian Nobel prize winning economist, Amartya Sen, theorizes that a famine is less likely to occur in a democracy because public pressure will ensure

---

93 UNDP. 2007. *Human Development Report 2007/2008*, p. 283.

active intervention.[94] However, there has been a constantly high level of chronic under-nourishment in over 60 years of democratic independence, and the press regularly reports on deaths due to starvation.

As a result of considerable effort, India has become self-sufficient in the production of key grains, and has an adequate per capita availability of food. The country has a strong legal framework in place for the protection of all human rights and has set up many programmes to address hunger and undernutrition, yet hunger and malnutrition persist. Those in need of food are unable to access it for reasons ranging from poverty, to lack of coherent strategies to combat hunger, to administrative inefficiency, discriminatory practices and weak governance.

2. Identifying the Hungry in india

India is engaging in an intensive debate with regard to identifying the poor and the hungry. The debate focuses on three main issues: (i) the definitions of starvation, hunger and food insecurity; (ii) the methodologies for defining the poverty line; and (iii) developments regarding the visibility of the situation of tribal communities and lower caste persons.

Defining Hunger and Starvation

The ICESCR makes a distinction between the right to adequate food and the fundamental right to be free from hunger. Freedom from hunger is frequently seen as meeting the 'minimum content' of the right to adequate food and requires immediate action, whereas starvation is an extreme form of hunger, which often leads to death.

The definition of hunger in India has been widely debated and is in the process of evolving from the idea of minimum caloric requirements to that of a balanced diet. Nutrition experts have defined the minimum caloric requirement in order to be free from hunger as 2 400 calories a day for rural adults (involved in physical labour) and 2 100 calories for urban adults.[95] This in turn has been calculated in terms of wheat and rice requirements. But dietary needs go beyond grains alone, so the daily requirement of protein and other nutrients have also been calculated. However, the focus so far has been on defining minimum food requirements rather than on looking at undernutrition outcome measures.

There has been discussion with regard to defining the entitlements of persons covered by the public distribution system (rice or wheat) and supplementary nutrition programmes, such as school feeding and food entitlements provided through the Integrated Child Development Service, where there is a prescribed amount of calories and proteins.

India's Famine Codes provide for damages to be paid to the family of a person who has died of starvation. However, the definition of starvation is very limited and does not include under-nourishment as a contributing factor. Consequently, if a person dies of tuberculosis while also being severely undernourished, that person will not be deemed to have died of starvation but of disease. Similarly, if an autopsy reveals food in a person's stomach, even as little as a grain of corn or a blade of grass, that person is deemed not to have starved to death. Hunger may have

---

94   Sen, A. 1981. *Poverty and Famine: An Essay on Entitlement and Deprivation*. Oxford. Clarendon Press.

95   Planning Commission. 2008. *11th Five Year Plan, 2007-2012*. Vol. II, p. 132.

driven people to eat grass or other substances unsuitable for human consumption, but the cause of death in such cases is described as food poisoning rather than starvation.

Post mortems do not prevent starvation, nor do they promote the right to food. On the contrary, a post mortem may even add to the indignities already suffered by the family of the deceased. In 2003, the National Human Rights Commission (NHRC) declared that it is wrong to insist on mortality as proof of starvation.[96] Rather, the continuum of distress should be the criterion and death from starvation should be prevented.

The dialogue on starvation needs to stop focusing solely on people who are dying of hunger, and move towards the inclusion of those who live with hunger as a way of life.

Undernutrition and Poverty in India

Progress in combating malnutrition has been slow. In 1998-99 an average of 36 percent of India's adult population had a body mass index below 18.5 (the threshold for adult malnutrition). Despite nearly a decade of robust economic growth, that percentage decreased only slightly in 2005-06 to an average of 33 percent, while the percentage of underweight children under three years of age remained unchanged (47 percent in 1998-99 and 46 percent in 2004-06)[97]. Table 1 provides an overview of some additional key child nutrition indicators.

The World Hunger Series report for 2006 highlights what it calls India's 'silent emergency' of chronic hunger and suffering. Each year, more people die from malnutrition than those who lost their lives (some 3 million) during the 1943 Bengal crisis – the last major famine in India. Far too many of these deaths are among children. India has just 17 percent of the world's population but 35 percent of all underweight children throughout the globe.[98]

Even among the richest 20 percent of the population, stunting rates stand at 25.3 percent for children under five,[99] which would suggest that malnutrition in India is not only a problem of poverty. Similarly, more than half of the children aged 6 to 59 months are anaemic, even if the mothers have had 12 or more years of education or belong to the highest wealth quintile.[100]

---

96   National Human Rights Commission. 2003. *Annual Report 2002-2003, Case No 37/3/97-LD, Orissa Starvation Deaths Proceedings, 17 January 2003*, p. 334.

97   Planning Commission. 2008. *11th Five Year Plan 2007-2012*, Vol. II. Social sector, p. 128.

98   WFP. 2006. *World Hunger Series 2006: Hunger and Learning*, p. 32. Rome.

99   International Institute for Population Sciences (IIPS) and Macro International. 2007. *National Family Health Survey (NFHS-3)*. 2005-06. Volume I, p. 271. Mumbai, India.

100 International Institute for Population Sciences (IIPS) and Macro International. 2007. *National Family Health Survey (NFHS-3)*, 2005-06. Volume I, p. 289. Mumbai, India.

Table 1. Child Nutrition Indicators	%
% of infants with low birth weight, 1999-2006	30
% of under-fives (2000-2006) who are underweight: moderate & severe	43
% of under-fives (2000-2006) who are underweight: severe	16
% of under-fives (2000-2006) suffering from wasting: moderate & severe	20
% of under-fives (2000-2006) suffering from stunting: moderate & severe	48

Source: UNICEF (Available at http://www.unicef.org/infobycountry/india_statistics.html).

While undernutrition rates are high everywhere in India, there are vast differences between one region and another: rates of underweight children under the age of five vary from 19.7 percent in Sikkim and 19.9 percent in Nagaland (in the Northeast), to 60 percent in Madhya Pradesh in the Centre. Bihar and Jharkand in the East also have very high rates of 55.9 percent and 56.5 percent, respectively.[101]

The number of people in India living below the national poverty line is indicated as being 28.6 percent.[102] However, international poverty lines would rate this percentage as being higher in reality, because 41.6 percent earn less than the purchasing power parity of US$ 1.25 a day (extreme poverty) and 75.6 percent earn less than US$ 2 a day (poverty).[103]

Definition of Poverty Line

India has set a national poverty line to serve as a benchmark for targeted interventions. The Below Poverty Line (BPL) category identifies people who need special care and protection; the Above Poverty Line (APL) category is also used and carries some lesser entitlements. The official poverty line is set at Indian Rupees (INR) 356.30 (US$ 7.40) per capita per month in rural areas and INR 538.60 (US$ 11) per capita per month in urban areas.[104] However, for the purpose of eligibility for assistance, BPL families are identified through extensive surveys carried out at the state level, using criteria developed by the Ministry of Rural Development.[105] The latest questionnaire sought information regarding the following:

- status of land and house ownership
- access to sanitation
- food availability
- ownership of consumer goods

---

101 International Institute for Population Sciences (IIPS) and Macro International, 2007. *National Family Health Survey (NFHS-3)*, 2005-06. Volume I, p. 273. Mumbai, India.

102 UNDP. 2008. *Human Development Report 2007/2008*. Reference years 1990-2004. New York.

103 World Bank. 2008. *World Development Indicators. Poverty Data*. Washington. The figures cited are from 2004-5.

104 Government of India. Press Information Bureau. 2007. *Poverty Estimates for 2004-05*. New Delhi, India. (Available at http://planningcommission.nic.in/news/prmar07.pdf).

105 Planning Commission. *11th Five Year Plan 2007-2012*, Vol. II, *Social sector*, p. 135.

- status of education in the family
- livelihood situation
- status of children[106]

The total number of people who can be classified as BPL is dependent on the state wide poverty estimates provided by the Planning Commission. The criteria for BPL status have been defined at a level that is criticized for being lower than the subsistence requirement. Consequently, therefore, activists argue that the 'poverty line' is closer to a 'starvation line'.[107] Following an intervention in the Supreme Court by the Right to Food Campaign, the Court ruled that the Government of India cannot reduce the percentage of people identified for benefits from the Public Distribution System (PDS) to less than 36 percent.[108]

In addition to the question of defining the poverty line, targeting of the PDS in 1997 also led to extensive inclusion and exclusion errors in the identification of beneficiaries for the programme. The Planning Commission estimates that not more than 57 percent of BPL families have BPL ration cards[109].

Some states, such as Chhattisgarh, have decided to expand the BPL category, using their own resources, to include also those who had been described as BPL in previous surveys that adopted different methodologies.

The poverty line definition has been criticized internally by the Commissioners of the Supreme Court,[110] among others, and externally by the Committee on Economic, Social and Cultural Rights, which has recommended that India review their national poverty threshold.[111]

A national commission examined unemployment in India and stated that some 77 percent of the population should be classified as both poor and vulnerable, with low daily expenditures, and that high economic growth benefited mainly the remaining 23 percent.[112]

Who Are the Most Vulnerable?

While poverty and under-nourishment levels are generally high, some groups and individuals are worse off than others. Destitution is more endemic amongst certain groups, such as people with disabilities, people with stigmatizing illnesses such as leprosy or HIV/AIDS, the elderly, the young who lack family support, and single women. Destitute people cut across Indian society and include the scheduled caste population, tribal populations, manual scavengers, beggars,

---

106 Jalan, Jyotsna and Murgai, Rinku. 2007. *An Effective 'Targeting Shortcut'?* An Assessment of the 2002 Below-Poverty Line Census Method.

107 Guruswamy, Mohan and Abraham, Ronald Joseph. 2006. *The poverty line is a starvation line*.
(Available at: http://infochangeindia.org/200610195662/Agenda/Hunger-Has-Fallen-Off-The-Map/The-poverty-line-is-a-starvation-line.html).

108 Supreme Court Order of 14 February 2006, *PUCL v Union of India and others*. Writ petition (civil) 196 of 2001.

109 Planning Commission. *11th Five Year Plan 2007-2011*, Vol. II, Social sector, p. 135.

110 Commissioners of the Supreme Court. 2007. *Seventh Report*.

111 Committee on Economic, Social and Cultural Rights. 8 August 2008. *Concluding observations on India*, par. 68. UN doc. E/C.12/IND/CO/5.

112 National Commission for Enterprises in the Unorganized Sector, Ministry of Small Scale Industries. 2009. *The Challenge of Employment in India*, p. iii.

sex workers, landless labourers and artisans. People who have been displaced by natural disasters or development projects are often included in this group.

The Supreme Court has identified the most vulnerable as being *'the aged, infirm, disabled, destitute women, destitute men in danger of starvation, pregnant and lactating women, and destitute children, especially in cases where they or members of their family do not have sufficient funds to provide food for them.'*[113]

In its preamble, the Constitution ensures gender equality as a fundamental right and also empowers the State to adopt positive discrimination measures in favour of women, through legislation and policies. In 1993, India ratified the Convention on Elimination of All Forms of Discrimination Against Women. Furthermore, over one million women have been elected to local *panchayats* as a result of a 1993 amendment to the Constitution requiring that one third of the elected seats to the local governing bodies be reserved for women.

Nevertheless, women face special nutritional risks because they are usually the last in the family to eat, and if food is scarce they are likely to eat the least. Also, women's literacy, education and earnings lag behind that of men. According to India's 11th Five Year Plan, these factors, along with the fact that women and girls are less likely than men to access health services in case of illness, contribute to the nation's skewed sex ratio (933 women to 1 000 men).[114]

More than one third of Indian women have a Body Mass Index below 18.5, which indicates a high level of nutrition deficiency. Nutrition problems are more serious for rural women, women with no education, scheduled tribe and scheduled caste women, and women in the bottom two wealth quintiles.[115]

One of the most prominent features of India's social structure is the caste system. There are about 3 000 castes. Of these, as many as 779 have been placed under the special protection of the Constitution as 'scheduled castes' because they are at the lower end of society. The scheduled castes were formerly known as 'untouchables' but generally refer to themselves as Dalits, which literally means 'oppressed'. The Indian Constitution of 1950 outlawed 'untouchability' and subsequent legislation has sought to provide further protection, in particular the Protection of Civil Rights Act (1955), and the Scheduled Castes and Scheduled Tribes (Prevention of Atrocities) Act (1989).

Nevertheless, the caste system remains one of the oldest institutionalized systems of oppression in the world. Dalits are much more likely than others to be poor and under-nourished. Figures from 2004-5 indicate that 36.80 percent of Dalits were living below the poverty line in rural areas compared to only 28.30 percent of other groups. In urban areas the gap was slightly larger: 39.20 percent of scheduled caste households were BPL compared to 25.70 percent among other households.[116]

---

113 Supreme Court Order of 2 May 2003. *PUCL v Union of India and others*. Writ petition (civil) 196 of 2001.

114 Planning Commission. *11th Five Year Plan 2007-2011*. Vol. II, Social sector, p. 132.

115 International Institute for Population Sciences (IIPS) and Macro International. 2007. *National Family Health Survey (NFHS-3), 2005-06*, Key Findings, p. 305. Mumbai, India.

116 Planning Commission. *11th Five Year Plan 2007-2012*, Volume I, pp. 106-107.

The scheduled caste and scheduled tribe populations constitute 16.2 percent and 8.2 percent respectively of the country's total population and both live mainly in rural areas. Whit respect to Dalits the proportion living in rural areas is close to 80 percent.[117]

The scheduled tribes, or Adivasis – India's indigenous people – also enjoy special protection under the Constitution. There are more than 700 of these tribes, totalling 84.3 million people (2001 figures). The majority live in the central and eastern parts of India and in the northeast. Many of the Adivasis lost access to their traditional land and are living mostly in the mountains, hills, forests, deserts or perpetually snow-bound areas. They have no access to public services, such as the Public Distribution System (PDS), and their geographical, social and political isolation frequently leaves them out of the reach of welfare schemes and social services.[118]

Children belonging to the scheduled castes, scheduled tribes, or other marginalized classes have relatively high levels of underweight, stunting and wasting. Children from scheduled tribes have the poorest nutritional status on almost every count, and the high prevalence of wasting in this group (28 percent) is particularly striking.[119]

Analysis of data from government surveys typically classified the population into income quintiles, which highlights factors related to poverty and inequality. However, this classification failed to identify groups such as Adivasis and Dalits who face social discrimination and are still frequently overlooked. The Commissioners of the Supreme Court have deplored the lack of official data but found plenty of anecdotal evidence for discrimination and social exclusion. They have urged that data collection methodologies be revised, so as to address the information gap and better identify the most vulnerable groups.[120] It is a positive development that the latest national family health survey and the government's latest five-year plan contained disaggregated data on the situation of Dalits and Adivasis.

---

117 Planning Commission. 2005. *Report of the Task Group on development of scheduled castes and scheduled tribes on selected agenda items of the national common minimum programme*, p. 66.

118 Housing and Land Rights Network & Habitat International Coalition. 2008. *The Human Rights to Adequate Housing and Land in India*. Parallel Report Submitted to the Committee on Economic, Social and Cultural Rights Under Consideration of Article 11.1 of the International Covenant on Economic, Social and Cultural Rights, p.6; and FIAN International. Parallel Report: The right to adequate food in India. Reference: Second to fifth periodic reports of India, UN doc. E/C.12/IND/5, submitted to the CESCR, 40th session, pp. 36, 39.

119 International Institute for Population Sciences (IIPS) and Macro International. 2007. *National Family Health Survey (NFHS-3), 2005-06*, Volume 1, p. 272. India.

120 Commissioners of the Supreme Court. 2007. *Seventh Report* (November 2007). The Eighth Report of the Commissioners (2008) focuses on a range of such marginalized groups and makes a case for specific entitlements for them. The communities taken into account in the report include the Adivasis, urban homeless, bonded labour and most discriminated Dalit groups.

3. Assessing Laws, Policies and Institutions

India has various institutions for assessing policies, laws and institutions related to food and nutrition security, such as the Planning Commission, when it prepares for the five-year plans. However, these assessments focus solely on the delivery and function of individual programmes and schemes and are not particularly focused on the right to food.

At the request of the Indian Government, the National Human Rights Commission is conducting a thorough assessment of the state of realization of the right to food that could serve as a basis for a National Action Plan on the Right to Food. The National Action Plan would incorporate current government programmes and seek to identify gaps and inconsistencies.

The UN Special Rapporteur on the Right to Food visited India in 2005 and submitted a report to the Commission on Human Rights.[121] Although the report is not a national right to food assessment, it contains an assessment of the legal, institutional and policy frameworks, as well as statistics on hunger and undernutrition, and an overview of direct violations of the right to food.

India, as a State Party to the International Covenant on Economic, Social and Cultural Rights (ICESCR), is obliged to submit periodic reports to the Committee on Economic, Social and Cultural Rights (CESCR). In 2007, it submitted its combined second, third, fourth and fifth periodic report providing selected statistics and descriptions of major initiatives and programmes regarding the right to food. These initiatives included not only schemes for the provision of food grains and farmer support but also efforts regarding land distribution.[122]

As part of the process of examining reports of States Parties to the ICESCR, FoodFirst Information and Action Network (FIAN) International submitted a parallel report on the right to food. This report contains detailed analyses and criticizes various aspects of India's legal and policy framework for the right to food.[123]

In the case entitled *PUCL v Union of India and others*, which became known as the **Right to Food Case**, the Supreme Court ordered[124] the establishment of a Central Vigilance Committee for the Public Distribution System (PDS), chaired by a retired Supreme Court judge and assisted by the Court's Commissioners in the right to food case. The Committee's mandate was to look into *'the maladies affecting the proper functioning of the Public Distribution System'* and to focus on (a) the method of appointing dealers; (b) the ideal commission or rates payable to the dealers; (c) ways in which the committees already in place can function better; and (d) how the allotment of food stock sold in the shops can be more transparent. The Committee issued its final report for the State of Delhi in August 2007[125] and made a number of detailed recommendations. It has subsequently reported on 12 states in India and is expected to report on the whole country.

---

121 UN. 2006. *Report of the Special Rapporteur on the Right to Food*, Jean Ziegler. *Addendum. Mission to India.* UN doc. E/CN.4/2006/44/Add.2.

122 Committee on Economic, Social and Cultural Rights. *Combined second, third, fourth and fifth periodic report of India.* UN doc E/C.12/IND/5, 01.03.2007.

123 FIAN International. 2008. Parallel Report: *The right to adequate food in India.* Reference: *Second to fifth periodic reports of India.* UN doc. E/C.12/IND/5, submitted to the CESCR, 40th session.

124 Supreme Court Order of 12 July 2006, *PUCL v Union of India and others.* Writ petition (civil) 196 of 2001.

125 Central Vigilance Committee for the Public Distribution System (PDS). 2007. Final report. (Available at: http://pdscvc.nic.in).

There have been thus a number of assessments of different aspects of the right to food undertaken by different actors. While none of these are national assessments as recommended by Right to Food Guideline 3, in scope or in process, there is certainly enough information available in India for such an assessment to take place, should the Government decide to adopt a national strategy on the right to food.

4. Sound Food Security Policy

The strategy followed by India is two-fold. On the one hand, it aims to increase the production of food grains and provide a minimum support price to farmers – which not only adds to the food grain stocks but also adds to farmers' income, thus increasing their individual purchasing power. On the other, it promotes interventions aimed at different population groups and at the distribution of food grains, cooked food and cash.

On the production side, India adopted high yielding varieties in the 1960s and 1970s and was part of the 'green revolution'.[126] The National Food Security Mission provides the current policy framework for the Department of Agriculture.[127] The section below will focus on selected points that might also be of interest to countries other than India.

Public Distribution System (PDS)

The Public Distribution System (PDS) has three objectives: 1) to ensure a minimum support price to farmers in support of their livelihoods; 2) to control the market price through stocking and controlled release; and 3) to make grain accessible to the poor through subsidized sales in fair price shops, food for work programmes and as free food under the Integrated Child Development Scheme (ICDS), the Mid-Day Meal Scheme (MDS), and others.

The PDS has evolved over time, from a universal system to a targeted one, so that only those who have BPL ration cards can buy subsidized grains.[128]

Following the introduction of the targeted PDS (in 1997), public grain stocks rose drastically between 1997 and 2001, since the targeting entailed less grains being distributed. However, stocks diminished between 2001 and 2007, during which period wheat stocks fell from 25.5 million tons to 4.5 million tons and rice stocks from 32 million tons to 13 million tons.[129]

According to India's Planning Commission, the PDS has failed to make grain accessible to the poor, as cereal consumption fell over the last two decades without being significantly replaced by other foods. The Commission identified the major deficiencies of the PDS as being: 1) widespread exclusion and inclusion errors; 2) non-viability of fair price shops; 3) failure to fulfil the price stabilization objective; and 4) leakages.[130]

---

126 See Ganguli, S. 2002. *From Bengali famine to green revolution*. Published in One stop India. (Available at http://www.indiaonestop.com/Greenrevolution.htm).

127 http://www.indg.in/agriculture/rural-employment-schemes/national-food-security-mission

128 Website of the Department of Food and Public Distribution. (Available at http://fcamin.nic.in/dfpd/EventDetails.asp?EventId=26&Section=PDS&ParentID=0&Parent=1&check=0).

129 Planning Commission. *11th Five Year Plan 2007-2012*, p. 133.

130 Planning Commission. *11th Five Year Plan 2007-2012*, p. 135. See also Programme Evaluation Organization, Planning Commission. 2005. *Performance Evaluation of the Targeted Public Distribution System*. PEO Report No. 189, p. ii. New Delhi.

Studies quoted in the *11ᵗʰ Five Year Plan* reveal that PDS leakage decreases when there is strong political commitment, careful monitoring by the authorities, awareness on the part of right holders, high literacy rates and active grass roots organizations.[131] The elimination of private retail outlets has also helped to stem leakage, as has been demonstrated in Chhattisgarh.[132]

Self-sufficiency

India has opted for self-sufficiency in basic staples, as a food security strategy. This implies support for production, and export controls to ensure that only surplus production is exported. The country is too large to be able to ensure food availability through international markets, since its needs are much greater than what is available through imports from other countries. It has produced sizeable surpluses in the past, with the result that world market prices were significantly influenced by Indian harvests and Indian export restrictions.

India reacted to the 2007-8 food price crisis by setting up export restrictions. It banned the export of all rice except *basmati*, with a view to keeping domestic prices stable and the justification of 'feeding its own first.' According to the World Bank Report on Global Development Finance, these policies have contributed to the increase in international grain prices and constitute a long-term disincentive for farmers to increase productivity and invest in agriculture.[133] The Special Rapporteur on the Right to Food recently reminded States Parties to the ICESCR of their international obligations with regard to food importing countries.[134]

Welfare Schemes

There is a plethora of schemes related to food and nutrition security in India – 80 by some counts. They have similar objectives and target populations but are executed by different agencies resulting in inefficiency, duplication of efforts, and gaps in coverage as implied by the following comment by the Planning Commission:

> 'The existing social security system in India exhibits diverse characteristics. There are a large number of schemes, administered by different agencies, each scheme designed for a specific purpose and target group of beneficiaries, floated as they are by the Central and State Governments as well as by VO [voluntary organizations] in response to their own perceptions of needs, of the particular time. The result is often ambiguous. Many a time some scheme(s) might be responsible for creating 'exclusion' of a large number of those 'in most critical need for support from the planning process,' on grounds of practicability or to protect the interests of those who are already 'in'. There are wide gaps in coverage (a large population is still uncovered under any scheme) and overlapping of benefits (a section of the population is covered under two or more schemes).'[135]

---

131 Planning Commission. *11ᵗʰ Five Year Plan 2007-2012*, p. 136.

132 Chhattisgarh. 2004. Public Distribution System (Control) Order.

133 The World Bank. 2008. *Global Development Finance 2008: The Role of international Banking*, p. 25. Washington, D.C.

134 UN. 2008. Report of the Special Rapporteur on the Right to Food, Olivier De Schutter: *Building resilience: a human rights framework for world food and nutrition security*, paras. 24-52. UN doc. A/HRC/9/23, 8 September 2008.

135 Planning Commission. *11ᵗʰ Five Year Plan 2007-2012*, p. 150.

Many of the benefits provided under the different schemes go to families rather than individuals. This can have discriminatory effects. For example, the PDS ration system, while providing additional subsidy for food grains, operates on the assumption that the average family consists of five people. This causes major problems for larger families and is also inconsistent with the basic notion that human rights are individual rights. The system may also work against women: while BPL cards should carry the names of all family members, there is anecdotal evidence to suggest that women have difficulty in accessing rations without the presence of their husbands. Following is a table giving an overview of the main schemes, and the entitlements provided.

Overview of Main Public Schemes for Food and Nutrition	
Scheme	Entitlements
Mid-Day Meal	Cooked school meal for all children in government funded primary schools in drought areas, also during vacation. Min. 450 calories and 12g of protein per meal.
Integrated Child Development Services (ICDS)	Take-home rations or cooked meals for all adolescent girls, pregnant and lactating women (500 calories and 20-25g of protein), and children under six (300 calories and 8-10g of protein) who come to the ICDS centre. Also, supplementary nutrition, nutrition education and more.
Targeted Public Distribution System (PDS)	Distributes food grains and other basic commodities at subsidized prices through fair price shops. BPL and APL families entitled to buy 35kg of grains per month, at different prices for BPL and APL.
National Maternity Benefit	One-time payment of INR 500 to pregnant women 8-12 weeks before delivery.
National Family Benefit	Lump sum of INR 10 000 to BPL families on the death of the primary breadwinner, defined as a person between the age of 18 and 65 whose earnings have contributed substantially to the family income.
National Old Age Pension Scheme	All BPL people over 65 are to receive INR 200 per month from the central government. States have been urged to provide an equal amount.
Antyodaya Anna Yojana (AAY)	AAY card holders (individuals and families identified as destitute) are entitled to 35kg of subsidized rice or wheat per month from the designated local ration shop. The subsidized price charged is INR 2 per kg for wheat and INR 3 per kg for rice.
Annapoorna	Indigent persons over 65, eligible for old age pension but not receiving it, receive 10kg of free grain every month.

Overview of Main Public Schemes for Food and Nutrition	
Scheme	Entitlements
Sampoorna Gramin Rozgar Yojana (Universal Food for Work Scheme)	Centrally-sponsored employment scheme, open to all rural poor who are in need of wage employment and desire to do manual and unskilled work in and around his/her village/habitat. Preference shall be given to agricultural wage earners, non-agricultural unskilled wage earners, marginal farmers, women, members of the scheduled castes/ scheduled tribes, parents of child labourers withdrawn from hazardous occupations, parents of handicapped children or adult children of handicapped parents who want to work for wage employment. This scheme is being automatically phased out in districts where NREGA has come into force (See section entitled 'Legal Framework to Realize the Right to Food').
Village Grain Bank	In drought prone areas, hot and cold desert areas, tribal areas, inaccessible hilly areas, entitlements linked to BPL/AAY entitlements.

India spends large amounts of human, organizational and financial resources on the PDS and the different schemes. Although there should be sufficient funds to ensure freedom from hunger and undernutrition at the very least, problems such as duplication, inefficiency, leakage, corruption, and others, are inhibiting progress. A systemic change is needed, as well as 'zero tolerance' of starvation and undernutrition. It is hard to say to which extent the strategies are insufficient and to which extent it is their implementation that is at fault.

5. Allocating Roles and Responsibilities

There are a number of important institutions in India contributing to the realization of the right to food. In addition to those agencies implementing up to 80 programmes and schemes aimed at availability and accessibility of adequate food, there are also institutions to ensure accountability, such as the Supreme Court and the National Human Rights Commission, which will be discussed below.

Special challenges regarding coordination are linked to the sheer size of the country: most of the constituent states would be considered large countries in their own right, in terms of inhabitants, had they been independent. Yet, despite India's food security challenges, it does not have a coordination body specifically in charge of food security and nutrition. There is a Planning Commission, which is a coordination body chaired by the Prime Minister, but it is responsible for all sectors.

The Ministry of Consumer Affairs, Food and Public Distribution is an essential player in the food entitlement schemes, providing food grains for distribution through the different schemes. The Food Corporation of India functions as an autonomous organization, working along commercial lines, to undertake purchase, storage, movement, transport, distribution and sale of food grains and other foodstuffs.[136] Other Ministries, such as the Ministry of Rural Development, may then be involved in the delivery of programmes. Central government contributions are frequently matched by those of state/union territory governments. Some states have successfully implemented centrally funded programmes while others have failed. The reasons for this vary and may be attributed to differences in resources, governance issues and culture.

The National Human Rights Commission has advocated on several occasions for *'convergence'* in the delivery of the various benefits, so that officials at the lowest levels would be responsible for the bundle of benefits and entitlements of a household.[137] Such convergence could improve both efficiency and coherence, and free up human resources for better monitoring and evaluation activities.

The lesson learned here is that unless there is proper coordination of policies and programmes and assurance that officials at the local level can effectively deliver assistance and be held accountable, there is bound to be overlapping and duplication of effort.

6. Legal Framework to Realize the Right to Food

Overall, India has a solid legal framework for the right to food: The constitutional protection of the right to food is strong, and legal provisions for the various entitlements are also strong, supported by the efforts of both Parliament and the Supreme Court, in addition to detailed Government orders. One aspect that needs to be improved, however, is institutional coordination. This could be covered by a framework law on the right to food or by a special right to food law.

Constitutional Provisions

The right to food is inherent in several provisions of the Indian Constitution, including the commitments in the preamble to secure 'social and economic justice' and 'equality of opportunity', supported by the commitment to promote the 'dignity of the individual.' The following articles in the Constitution, when read together, highlight the State's obligation to ensure food security as an entitlement.

Article 21, in Part III on fundamental rights, states: *'No person shall be deprived of his life or personal liberty, except according to procedure established by law.'* However, the term 'life' in this article has been judicially interpreted to mean the right to live with human dignity, and not merely to survive or exist. In the words of the Supreme Court: *'Right to life includes the right to live with human dignity and all that goes along with it, namely the bare necessaries of life such as adequate nutrition, clothing and shelter.'*[138] It thus includes all aspects which would make life meaningful and complete.

Article 39 (a), in Part IV, on Directive Principles of State Policy, provides that 'the State shall, in particular, direct its policy towards securing (a) that the citizens, men and women equally, have the right to an adequate means of livelihood.' Furthermore Article 47, also in Part IV,

---

136 Programme Evaluation Organization. Planning Commission. 2005. *Performance Evaluation of the Targeted Public Distribution System*. PEO Report No. 189, p. 2. New Delhi, India.

137 See, for example, National Human Rights Commission. *Annual Report 2004-5*, p. 134.

138 Supreme Court of India. 1981. *Francis Coralie Mullin vs The Administrator, Union Territory of Delhi* (AIR 1981 SC 746 (1981) 1 SCC 608, 1981 LJ 306).

states: *'The State shall consider raising the level of nutrition and the standard of living of its people and the improvement of public health as among its primary duties.'*

The Supreme Court has given fuller meaning to the right to life provision by referring to the provisions of Part IV, which were originally intended not to give rise to individual claims. According to the National Human Rights Commission, the reading of Article 21 together with Articles 39 (a) and 47 *'places the issue of food security in the correct perspective,'* making the right to food a guaranteed fundamental right, which is enforceable by the Supreme Court by virtue of the remedy provided under Article 32 of the Constitution. In light of the foregoing, the State is obliged to provide for minimum livelihood needs, standards of living and level of nutrition.[139]

The Constitution thus gives legal effect to Article 11 of the ICESCR, which was ratified by India in 1979. The lesson to be drawn from this is that the right to food can be justiciable even without explicitly justiciable constitutional provisions.

Legal Framework for PDS

The *Essential Commodities Act* of 1955 provides the legal basis for the PDS, now known as the Targeted PDS. It provides for the control and regulation of the production, manufacture and distribution of essential commodities in the country, for the general public good. It gives the State considerable authority to regulate and control the operations of private actors, so as to control market vagaries.

When there was severe drought in 1980-81, the *Essential Commodities (Special Provisions) Act (ECA)* of 1981 was enacted *'for a temporary period for dealing more effectively with persons indulging in hoarding and black marketing of, and profiteering in, essential commodities and with the evil of vicious inflationary prices and for matters connected therewith or incidental thereto.'*[140] This act can be invoked in times of crisis. This was done, for example, when the price of onions became exorbitant. This ECA overlaps the *Prevention of Black Marketing and Maintenance of Supplies Act* of 1980. The Act permits detention in certain cases to prevent black marketing and the maintenance of supplies of essential commodities to the community.

The *PDS (Control) Order* of 2001 covers a range of areas relating to correct identification of BPL families, the issuing of ration cards, and the proper distribution and monitoring of PDS-related operations. Breaching of the order is punishable under the ECA.

The central government issued a *Citizen's Charter* in 1997 (revised in July 2007), for adoption and implementation by state governments, in relation to increased transparency, accountability and participation in the PDS system. It contains, *inter alia*, basic information of interest to consumers and a model procedure and time schedule for the services. The Charter contains also information on the entitlements of BPL families, fair average quality of food grains, information regarding fair price shops, procedures for issuing ration cards, inspection and checking, right to information, vigilance, and public participation. However, these charters have not caught the public eye to the extent that their potential might suggest.

---

139 National Human Rights Commission. *Annual Report 2002-2003. Case No 37/3/97-LD. Extract from the proceedings of the Commission held on 17 January 2003 in relation to allegation of starvation deaths in KBK districts of Orissa*, p. 333.

140 http://business.gov.in/starting_business/infrastructure.php

Essential commodities are regulated more strictly than in most other modern states, and there are detailed legal provisions governing the PDS. However, these have not sufficed to obviate the major faults in the system, discussed above. Deregulation of the grains market has been suggested by economists since 1991 but has never been politically feasible.

Right to Information

The *Right to Information Act* 2005 gives every citizen the freedom to secure access to information under the control of public authorities, consistent with public interest. This is in order to promote openness, transparency and accountability in administration and in relation to matters connected therewith or incidental thereto.

Under this act, public authorities are obliged to provide information upon request and also to ensure, proactively, that people are informed of important policies, plans and programmes. Civil society organizations have successfully used the Freedom of Information Act to identify irregularities in the implementation of the PDS and force local authorities to open their books for scrutiny.[141]

The Right to Information Act has had an enormous impact on governance in the country and is already having an enduring impact on the battle against corruption at all levels. It has provided a tool for action, not only for activists and NGOs, but also for common people at large who are invoking its provisions for improving civic life and gaining access to public services in an unprecedented manner.

Employment Guarantee

The National Rural Employment Guarantee Act (NREGA) 2005 is a law whereby any adult living in a rural area, and willing to undertake unskilled manual work at the minimum wage is entitled to be employed on local public works within 15 days of applying. If employment is not provided within 15 days, the applicant is entitled to an unemployment allowance of at least a quarter of the minimum wage for the first 30 days and at least half the minimum wage thereafter. Each nuclear family is considered a household under the Act, and each is entitled to 100 days of employment per year. The work should be provided within 5 km of residence, or else a travel allowance must be paid.

Labourers are entitled to the statutory minimum wage for agricultural labourers in each state and wages are to be paid directly to the worker, weekly, in front of the community. The law also prescribes mandatory facilities such as drinking water, shade, medical aid and a crèche, if there are more than five children below the age of six, to be provided by the implementing agency.

There are special provisions to protect women's rights against discrimination of any kind, and equal wages must be paid to men and women alike. Priority should be given to women in the allocation of work and at least 33 percent of labourers should be women.

All documents related to implementation of the law must be available for public scrutiny; copies of documents must be provided at a nominal cost and muster rolls shall be proactively displayed in the administrative building of each *panchayat*, i.e. governmental administrative unit traditionally created at village level. In addition, the law provides for social audits by

---

141 See, for example, Commissioners of the Supreme Court. *Seventh Report, 2007*, p. 156.

village assemblies, and the provision of all relevant documents by the village council and other implementing agencies.

The Right to Food Campaign India was active in the promotion of this Act (NREGA), and sees it as one of its major victories.[142] It was also one of the rare examples of both State and civil society working closely together on pro-poor legislation with intense public involvement at every stage of its development.

Implementation of NREGA has met with some difficulties, in particular lack of administrative and technical manpower to ensure that its provisions are followed, especially the transparency measures. This has made verification of delivery to right holders difficult. Government audits have also found numerous instances of re-routing, misutilization and delays in transfer of funds.[143]

The enactment of NREGA highlights the strong relationship between the right to food and the right to work. It is also ground breaking from a human rights legal perspective, as it brought the safety net measure of public works squarely within the sphere of the rule of law and legal empowerment of the poor. Therefore NREGA is one of India's major achievements and can be considered a milestone in the implementation of the right to food in the country.

Relief Codes

Disaster relief is the responsibility of individual states, but the National Disaster Management Authority is the apex body at the national level, mandated by the Government of India to set out the policies, plans and guidelines on disaster management that will ensure timely and effective response to disasters.[144] The institutional management mechanism for disaster management at the state level is based on Relief Codes (in some states known as Scarcity Codes, Famine Codes, Relief Manuals, etc.), which focus on the relief aspect of disaster management. These documents are updated from time to time, but their basic framework can be traced back to a 1910 Model Famine Code of the British colonial administration.[145]

The Relief Codes are invoked mainly in times of scarcity and give detailed descriptions of the circumstances in which a famine could be declared, the measurement of the intensity of the famine and the relief provisions that need to be undertaken in such cases. The Relief Codes generally provide for specific entitlements of certain amounts of grains and proteins per day. These are provided as food for work for able-bodied adults and 'gratuitous relief' for those unable to work. The Supreme Court has declared the Relief Codes to be binding on the relevant state – unless there are better measures available – thus creating legally binding entitlements.[146]

Their more controversial provisions concern official accountability and payment of damages in the case of starvation deaths. The narrow definition of starvation and the links to accountability have

---

142 See the website of the campaign (Available at http://www.righttofoodindia.org/rtowork/ega_intro.html).

143 CAG (Comptroller and Auditor General). 2007. Draft Performance Audit of Implementation of NREGA, p. 95, quoted in Datt, Ruddar, 2008. *Dismal Experience of NREGA: Lessons for the Future*. 2008. Vol XLVI, No 17.

144 http://ndma.gov.in/wps/portal/NDMAPortal

145 National Human Rights Commission. Annual Report 2002-2003. *Case No 37/3/97-LD. Extract from the proceedings of the Commission held on 17 January 2003 in relation to allegation of starvation deaths in KBK districts of Orissa*, p.333.

146 Supreme Court Order of 2 May 2003. *PUCL v. Union of India and others*. Writ petition (civil) 196 of 2001.

caused deep denial at state levels, where every effort is made not to recognize starvation deaths, although there is no such problem with regard to recognizing undernutrition.

The National Human Rights Commission has discussed the Relief Code of the State of Orissa and concluded that it should be revised so as to accomplish a paradigm shift from the 'benevolence' domain to that of 'right'; to change from assessing harvest for their interventions, to assessing hunger; to shift the timing of support so as to include the hunger season and to devise terms for cognizance of starvation and destitution, other than medical autopsies of starvation deaths.[147]

The Ministry of Home Affairs has suggested that states amend the existing Relief Codes/Manuals into comprehensive Disaster Management Codes or Manuals, incorporating the aspects of preparedness, mitigation and planning measures at all levels.[148] It can be assumed, therefore, that these Relief Codes will soon be of historical importance only. However, they have played a significant role in India's debate on the right to food since 1996.

Draft Food Security Bill

The Congress party included a right to food act in its election manifesto for the 2009 parliamentary elections drawing on the perceived success of the NREGA. The government initiated the drafting of a food security bill and the Empowered Group of MInisters for Food was given the task to outline the framework of the bill. Civil society – the Right to Food Campaign in particular – followed the process closely and made various demands for a more comprehensive coverage of benefits and beneficiaries. The bill itself did not establish a new institutional framework for coordination or define the right to food but focused on some specific entitlements for the poor and food insecure. It centered around the PDS and mostly on relevant entitlements under scrutiny by the Supereme Court. The debate is expected to last well into 2010 and perhaps beyond.

7. Monitoring the Right to Food

Monitoring at the technical level is carried out by numerous Government departments, as well as the Planning Commission. Human rights focused monitoring is carried out mainly by the National Human Rights Commission and the Commissioners of the Supreme Court.

National Human Rights Commission

The National Human Rights Commission (NHRC) was established by the *Protection of Human Rights Act, 1993*, and was the first institution of its type to be established in South Asia. The mandate and independence of the NHRC are in compliance with the UN Paris Principles relating to the status and functioning of national institutions for the protection and promotion of human rights.[149]

The above Act stipulates that membership of the NHRC, with emphasis on drawing on the experience of senior judges, shall consist of five persons, three from the judiciary and two with

---

147 National Human Rights Commission. Annual Report 2002-2003. *Case No 37/3/97-LD. Extract from the proceedings of the Commission held on 17 January 2003 in relation to allegation of starvation deaths in KBK districts of Orissa*, pp. 334-5.

148 Ministry of Home Affairs, National Disaster Management Division. 2004. *Disaster management in India: A status report*, p.12. (Available at: http://unpan1.un.org/intradoc/groups/public/documents/APCITY/UNPAN029426.pdf).

149 Adopted by UN General Assembly. *Resolution 48/134 of 20 December 1993*.

practical experience in human rights. In addition, the Chairpersons of the National Commissions for Minorities, Scheduled Castes, Scheduled Tribes and Women, are all deemed to be members of the NHRC for discharging functions other than inquiries related to human rights violations or neglect by public officials.

The NHRC has a very heavy caseload: between 1 April 2004 and 31 March 2005 it received 74 401 complaints and dealt a total of 85 661.[150] Individual states have also begun to establish State Human Rights Commissions. By May 2009, 18 such commissions had been set up out of 35 states and union territories.[151] As these commissions become functional, they should be able to lighten the NHRC caseload.

The NHRC has been active with regard to the right to food and has investigated complaints about ongoing starvation in Orissa. The Commission often monitors a particular situation for an extended period of time, requesting quarterly performance appraisals related to the short- and long-term achievements of physical and financial targets. This was what took place regarding the Districts of Kalahandi, Balangir and Koraput (KBK) in the State of Orissa.[152] A Special Rapporteur was appointed for this case, as well as for cases of farmer suicides in Andhra Pradesh, Kerala and Karnataka. A section of the NHRC annual report is devoted to food security.[153]

The NHRC constituted a Core Group on the Right to Food in January 2006, composed of experts from across the country who have contributed to the right to food debate. The Core Group provides advice to the NHRC and has recommended the drawing up of a National Plan of Action.

The work of the NHRC is remarkable for its long-term engagement with a state on a recurring human rights issue, that is, the right to food, and also for the analysis of the Famine Codes, which were found to be inconsistent in some ways with the right to food.

Commissioners of the Supreme Court

In the ongoing Right to Food case, the Supreme Court appointed Commissioners for the purpose of monitoring the implementation of the Court's orders.[154] The Commissioners are empowered to enquire about any violations of these orders and to demand redress, with the full authority of the Supreme Court. They also report to the Supreme Court from time to time and may seek interventions going beyond existing orders, if required. The Commissioners are also empowered to monitor and report to the Court on implementation by the central and State governments of the various welfare measures and schemes.[155]

The Commissioners present periodic reports to the Supreme Court. These typically deal, first and foremost, with the implementation of Supreme Court orders. In addition, they attempt to highlight issues that need further directions from the Court. The reports are based on extensive correspondence with State governments, reports from the commissioners' advisors,

---

150 National Human Rights Commission. *Annual Report 2004-2005*, p. 2.

151 National Human Rights Commission's website. (Available at: http://nhrc.nic.in).

152 National Human Rights Commission. *Annual Report 2002-2003*, p. 197.

153 National Human Rights Commission. *Annual Reports 2002-2003 and 2004-2005*.

154 Supreme Court Order of 8 May 2002, *PUCL v. Union of India and others*. Writ petition (civil) 196 of 2001.

155 Supreme Court Order of 29 October 2002, *PUCL v. Union of India and others*. Writ petition (civil) 196 of 2001.

interaction with citizens' organizations, and field visits made by the Commissioners. So far, eight reports have been submitted, as well as some interim reports. They are a rich source of information on the food situation in India, on the implementation of interim orders and the functioning of various schemes. The reports also include detailed recommendations to the Supreme Court.[156]

The Commissioners also undertake field visits to the states to make grassroots assessments and enforce compliance of Supreme Court orders, as well as to engage and negotiate with the state governments. They are represented in both civil society groups and government policy formulation bodies with regard to right to food issues and have created a space for strong state-civil society interface. In fact, they constitute a unique mechanism in the world for monitoring the right to food.

8. Legal and Administrative Recourse Mechanisms

The Supreme Court of India may be described as an 'activist' court. It has not shied away from adapting the interpretation of the provisions of the Constitution to the needs of the times, thus effectively reaching its social justice objectives.

For example, under the Constitution, international treaties and conventions ratified by India do not automatically acquire the status of law. Implementation legislation must be adopted by Parliament and then incorporated into domestic law. However, the Supreme Court has held that international human rights conventions are to be read into Indian law.[157]

The Supreme Court also developed a doctrine allowing public interest litigation. Any person or group can file a writ petition before the Supreme Court and the High Court for the enforcement of the fundamental rights of the poor, the illiterate and the oppressed.[158] Further innovation lies in the role of the Supreme Court when such public interest litigation has been successfully filed. The Court considers itself to have a duty to gather further evidence, through the appointment of commissioners or other means. Moreover, the burden of proof shifts to the State which thereby has to demonstrate that its actions are legal.[159]

Finally, the Supreme Court holds that when it comes to the enforcement of human rights, it will not entertain notions from the State as to its financial resources, which it refers to as 'perverse expenditure logic'.[160] There have been hundreds of public interest litigations before the Supreme Court for the enforcement of economic, social and cultural rights. The focus in this study will be on the most famous case regarding right to food.

---

156 Refer to Commissioners' website for details (Available at http://www.sccommissioners.org).

157 Gonsalves, C. 2007. *From international to domestic law: the case of the Indian Supreme Court in response to ESC rights and the right to food*, in Wenche Barth Eide and Uwe Kracht. 2009. *Food and human rights in development. Vol. II: Evolving issues and emerging applications.* Antwerp-Oxford, p. 217.

158 Gonsalves, C., p. 218.

159 *Ibidem*, p. 219.

160 *Ibidem*, p. 220.

PUCL v. Union of India and Others (Right to Food Case)

The People's Union for Civil Liberties (PUCL), Rajasthan, commenced public interest litigation in April 2001 before the Supreme Court of India, following a number of starvation deaths in the State of Rajasthan which occurred while government warehouses had an abundance of grain stockpiled. The case was subsequently extended to cover the entire territory of the Union of India and all its states and union territories. The Supreme Court stated the following, regarding this case:

> 'In this petition... various issues have been framed many of which may have a direct and important relevance to the very existence of poor people: their right to life and the right to food of those who can ill-afford to provide their families with two meals a day. ...what is of utmost importance is to see that food is provided to the aged, infirm, disabled, destitute women, destitute men who are in danger of starvation, pregnant and lactating women and destitute children, especially in cases where they or members of their family do not have sufficient funds to provide food for them. In case of famine, there may be shortage of food, but here the situation is that amongst plenty there is scarcity. Plenty of food is available, but distribution of same amongst the very poor and the destitute is scarce and non-existent, leading to malnourishment, starvation and other related problems.' [161]

The basic argument of the PUCL was that the right to food is part of the fundamental 'right to life' enshrined in Art. 21 of the Indian Constitution. The plaintiff requested enforcement of the various food schemes and the Famine Codes (permitting the release of grain stocks in times of famine).

Various interim orders were made by the court over ten years; it has interpreted the constitutional right to life in light of the directive principles concerning the State's duty to raise the level of nutrition and the standard of living of its people. It has found that the prevention of hunger and starvation 'is one of the prime responsibilities of the government – whether central or state.' [162] Early on in the case, the court converted the benefits of eight food-related schemes into 'legal entitlements' by interim order and directed state governments to fully implement these schemes as per official guidelines. [163] These were:

• Integrated Child Development Services
• Mid-Day Meal Scheme
• Targeted Public Distribution System
• Antyodaya Anna Yojana (highly subsidized grains for the destitute)
• Sampoorna Gramin Rozgar Yojana (food for work scheme)
• National Old Age Pension Scheme
• National Family Benefit Scheme
• Annapoorna Yojana (free food grain for destitute poor over 65)

The court has ordered that the Famine Code of Rajasthan be implemented for three months; that grain allocation for the food for work scheme be doubled and financial support for schemes

---

161 Supreme Court Order of 2 May 2003, *PUCL v Union of India and others*. Writ petition (civil) 196 of 2001.

162 Supreme Court Order of 20 August 2001, *PUCL v Union of India and others*. Writ petition (civil) 196 of 2001.

163 Supreme Court Order of 28 November 2001, *PUCL v Union of India and others*. Writ petition (civil) 196 of 2001.

be increased; ration shop licensees must stay open and provide grain to families below the poverty line at the set price; BPL families' rights to grain must be publicised; all individuals without means of support (elderly people, widows, disabled adults) are to be granted an *Antyodaya Anna Yojana* ration card for grain; and state governments should introduce one hot meal per day in schools. In addition, modifications to the National Maternity Benefits Schemes have been suggested. The court has paid special attention to the situation of Dalits in the implementation of the Mid-Day Meal Scheme, to the effect that all children must eat together and that cooks should also come from the Dalit caste.[164]

It is interesting to note that the Supreme Court did not merely direct the states to formulate appropriate schemes for food distribution, as was done earlier in several cases, but went several steps further in directing strict implementation of already formulated schemes within fixed time frames, to make them entitlements and to ensure accountability. The Court's detailed instructions in interim orders have served as legislation and perhaps replaced the need for parliament to enact a legal framework around the various entitlements.

The orders of the Supreme Court in the right to food case are of historic significance in the economic rights discourse. Far from being an abstract example, the case has had a real impact on the ground. The universalization of the Mid-Day Meal Scheme, whereby every single child in the country attending primary school is provided with a school lunch, is one example: 120 million children are currently covered by this programme. There has been a similar effect on the Integrated Child Development Services (ICDS) programme which can potentially address the supplementary nutritional needs, pre-school education, immunization and health care needs of 160 million children across the country.[165]

Orissa Starvation Case

The National Human Rights Commission (NHRC) acted on reports of many instances of starvation in the State of Orissa, particularly the KBK (Kalahandi, Balangir, Koraput) Districts, on the basis of an individual complaint, as well as through remittance from India's Supreme Court.[166] The situation in the KBK Districts was particularly severe: these areas are notorious for hunger and starvation. The NHRC held a number of hearings in what became known as the Orissa Starvation Case, from 1997 to 2006 (when the case was officially closed) and issued specific orders in 1998 and 2003.

The NHRC set out short and long-term measures for ameliorating suffering and bringing relief to the people. The situation in the KBK Districts was monitored through reports from a Special Rapporteur of the NHRC as well as quarterly progress reports received from the State government. A comprehensive and convergent approach was recommended and adopted to cover different areas which, directly and indirectly, affect the right to food. These include social security schemes such as old age, widow and disability pensions, the Emergency Feeding Programme, supplementary nutrition programmes, the PDS and the National Family Benefit Scheme, all of

---

164 See http://www.righttofoodindia.org/orders/interimorders.html for summaries and full texts of the various orders.

165 See Commissioners of the Supreme Court. 2007. *Seventh Report.*

166 Writ petition (civil) No. 42/97 filed by the Indian Council of Legal Aid and Advice alleging that starvation deaths continued to occur in certain districts of the State of Orissa (Koraput, Balangir and Kalahandi – KBK districts).

which were monitored. Attention was also paid to health care, water supply and employment generating schemes.[167] As part of the same case, the NHRC also discussed Orissa's Relief Code and the nature of the right to food. It recommended that the Code undergo a revision process to ensure its consistency with human rights.[168]

Administrative Recourse for Implementation of Schemes

Administrative recourse is integral to the legal system in India, and people are entitled to lodge complaints with authorities at all levels. However, social exclusion and lack of political power may negate in practice the rights that people have on paper.

In addition, the Right to Food Case has given rise to legislative efforts by the Supreme Court on a grievance redress mechanism.[169]

The *Gram Sabhas* (village assemblies) are empowered to conduct a social audit on all food and employment schemes and to report all instances of misuse of funds to the respective implementing authorities who, on receipt of such complaints, shall investigate and take appropriate action in accordance with law. The Supreme Court specified the duties of the Chief Executive Officers/ District or *Panchayat* Collectors with regard to receiving, handling and solving complaints. Finally, the Commissioners of the Supreme Court are empowered to recommend corrective actions, upon which state or union territory administrations must act and report compliance.

9. Capacity Building – The Essential Role of Civil Society

The Right to Food Campaign in India is closely related to the PUCL petition before the Supreme Court but it has now grown much beyond the court case. It is an informal network of nearly 1500 organizations across the country, including trade unions, people's movements, NGOs, women's groups and networks, as well as individuals committed to the right to food. The work of the Campaign covers advocacy for legislative action, enforcing accountability through local action and ensuring the implementation of Supreme Court orders. As a decentralized network, it builds on local initiatives and voluntary cooperation. The Campaign has a very small secretariat, manned by volunteers.[170]

The Right to Food Campaign has taken up a wide range of aspects of the right to food. Its main demands include:

* A national Employment Guarantee Act
* Universal mid-day meals in primary schools
* Universalization of the Integrated Child Development Services (ICDS) for children under the age of six
* Effective implementation of all nutrition-related schemes
* Revival and universalization of the public distribution system

---

167 National Human Rights Commission. *Annual Report 2004-2005*, p. 94.

168 National Human Rights Commission. 2003. *Orissa Starvation Death Proceedings dated 17 January 2003*. Also reported in *Annual Report 2002-2003*. Annex 18.

169 Supreme Court Order of 8 May 2002. *PUCL v Union of India and others*. Writ petition (civil) 196 of 2001.

170 http://www.righttofoodindia.org/index.html

- Social security arrangements for those who are not able to work
- Equitable land rights and forest rights[171]

Some of these demands have already been met to a certain extent. For example, the National Rural Employment Guarantee Act was adopted in August 2005, and cooked mid-day meals have been introduced in all primary schools. The Planning Commission supports the universalization of the ICDS.[172] Further issues are likely to be taken up as the campaign grows.

The campaign began with the PUCL Right to Food Case before the Supreme Court. As mentioned earlier, the public interest litigation was successful. By December 2009, the case had still not closed; however it soon became clear that the legal process, on its own, would not go very far. This motivated the effort to build a larger public campaign for the right to food.

Consequently, a wide range of activities have been undertaken to further the demands of the Campaign. Examples include public hearings, rallies, picketing, marches, conventions, action-oriented research, media advocacy, and lobbying of Members of Parliament. One example is that on 9 April 2002 activities of this kind took place across the country as part of a national 'day of action on mid-day meals'. This event was instrumental in persuading several state governments to initiate cooked mid-day meals in primary schools. Similarly, in May and June 2005, the Campaign played a leading role in the '*Rozgar Adhikar Yatra*,' a 50-day tour of India's poorest districts, to demand the immediate enactment of a National Employment Guarantee Act. Three national conventions have been held so far – in Bhopal in June 2004, in Kolkata in November 2005 and in Bodhgaya in April 2007.[173]

10. Conclusions

India is a country where the right to food is far from being realized for all. Despite economic growth and policies aimed to ensure food availability and access by the poor, hunger is still widespread. Some social and ethnic groups are more vulnerable than others, in particular Adivasis and Dalits, while women's social status remains low, despite legal reform.

On the other hand, India is among the countries where there has been most in-depth discussion on right to food, at least within legal and non-governmental circles. The use of public interest litigation for the right to food is one way of ensuring access to justice for the poor. The Supreme Court has not only acknowledged that the right to food is a fundamental right under the Indian Constitution, but has issued detailed orders of a quasi-legislative nature. This is of the utmost importance: it sets an example for lawyers and judges worldwide to be creative and work within the legal system of a country while taking a leadership role in promoting the progressive realization of the right to food in the country concerned.

Long-term engagement with the right to food is integral to the methodology of the Supreme Court. The Right to Food Case has been ongoing since 2001 and up to the time of reviewing this paper (December 2009), the Supreme Court shows no signs of issuing a final judgment. This has enabled the Court, with the support of its Commissioners, to keep the issue under active

---

171 http://www.righttofoodindia.org/campaign/campaign.html

172 Planning Commission. *11ᵗʰ Five Year Plan*. Vol. 2, p. 141.

173 http://www.righttofoodindia.org/campaign/campaign.html

and constant review – another innovative way of bringing together monitoring and recourse. Similarly, the National Human Rights Commission investigated and followed up on starvation in the State of Orissa for a full decade – from 1997 to 2006 – fielding special rapporteurs and demanding quarterly progress reports from the government on suggested reforms.

While framework laws on the right to food have been adopted in a number of countries, India opted for detailed legislation on specific social programmes, through parliament and Supreme Court orders. Thus, school children's entitlements with regard to cooked mid-day meals are very precise, as are the entitlements of rural households to guaranteed employment in public works for a minimum wage.

Still, there are some lessons learned from India that are important to note even as the debate continues. It is clear that legal action with *strategic* objectives can have a great impact on the ground, particularly if linked to public campaigning. Aside from the Right to Food Campaign, India has a variety of schemes under way for promoting food security and nutrition. But considering the size of the country, it is not easy to coordinate these schemes and ensure their coherence. The main problems encountered relate both to the design of the schemes in place as well as to duplication, inefficiency, leakage and corruption – all of which are hindering the implementation of the different programmes. With proper coordination of policies and programmes, convergence of delivery of services and entitlements at local levels, and stronger accountability mechanisms, the fight against hunger and malnutrition in India could be much more successful. It is also important to note that in the case of India, the constitutional protection of right to food is strong and inherent in several constitutional provisions. In fact, the Supreme Court has included the right to food under the right to life in Article 21, thus giving it constitutional enforceability. Therefore, while parliaments play a key role in promoting policies and adopting legislation on social programmes that guarantee the right to food, courts also perform the paramount function of protecting the right through a wider interpretation of other fundamental rights.

Recommendations

- Legal action is strongly recommended as a tool particularly when coupled with a strategic objective to promote the right to food and when linked to public advocacy campaigns.
- Where there are a variety of schemes to promote food security, it is key to find a mechanism for coordination of schemes to ensure coherence in policy making and implementation.
- A careful analysis of programme design and examination of the risks of duplication, inefficiency, leakage and corruption should be priority concerns that guide the implementation of right to food policies.
- Coordination of policies and programmes is best in the form of convergence in the delivery of services and entitlements at the local levels with accountability mechanisms to ensure transparent implementation processes.
- Constitutional enforceability of the right to food is recommended as a strong form of protection of rights. But when the right to food is not listed as an explicit constitutional right, it can still be protected by an active Court's interpretation of fundamental rights in the Constitution that inherently protect the right to food.

V. MOZAMBIQUE
Fighting Hunger with a Human Rights Based Approach

Main Points

- Like many sub-Saharan countries, Mozambique is fighting the 'triple threat' of poverty, hunger and HIV/AIDS. The Government has taken up the challenge and is following a human rights based approach to tackle hunger.
- The President of Mozambique is leading a national, multi-sectoral programme aimed at implementing a green revolution to end hunger and poverty, and to create jobs, especially in the districts concerned. This is intended to be a holistic approach, highlighting the right of all people to have access to food that is culturally acceptable, nutritionally adequate, and sufficient in quantity for people to lead a healthy and active life. The Government sees food as a fundamental human right and not as charity.
- A manifestation of the Government's pledge to realize the right to food is the implementation of the country's Food Security and Nutrition Strategy, ESAN II, launched in 2008.
- In order for food and nutrition programmes to be effective there has to be a law in existence to enforce them. The Ministries of Agriculture and Justice expect to have a draft proposal for a Right to Food Law ready by 2010. The process is both consultative and participatory, and will raise awareness among duty bearers and right holders alike as to the content of the legislation.

1. Background

Although Mozambique has not yet ratified the International Covenant on Economic, Social and Cultural Rights (ICESCR), the country is advanced in implementing the right to food. Since the start of the new Millennium, an increasing number of civil society and government players are actively promoting the application of a human rights based approach to combat hunger. The Government sees food as a fundamental human right, not as charity, and is using social networks to reach the most vulnerable groups. It is also trying to improve the early warning system in order to minimize the effects of natural disasters.

When Mozambique became independent in 1975, after a long and devastating civil war, food insecurity and poverty were pervasive. The country had suffered more than 20 years of conflict[174] with brutal consequences for the population and the infrastructure to the point that for quite some time the country was known as the most food insecure country in the world.[175]

---

174 Following an armed conflict for independence between the guerilla forces of the Liberation Front of Mozambique (FRELIMO) and Portugal, lasting for almost 11 years and resulting in a negotiated independence in 1975 (1964-1975), a 12-year civil war broke out (1980-1992) between the FRELIMO and the Mozambican National Resistance Movement (RENAMO).

175 FAO. 2002. *The State of Food Insecurity in the World 2002*. Rome.

Since the signing of the General Peace Agreement in Rome in 1992, Mozambique has experienced rapid economic growth, averaging annual increases of 6.3 percent per capita GDP from 1996 to 2003 and 7.5 percent from 2003 to 2006. It has also experienced a considerable reduction in poverty levels which dropped from 69 percent of the population living below the poverty line in 1997 to 54 percent in 2003 (PARPA II, 2006).[176] In addition, the economic forecast for now suggests a growth rate of 6.5 percent per year between 2010 and 2014.[177]

However, in spite of the economic development, there have not been any significant improvements in food security and nutrition; Mozambique still ranks among the ten countries with the highest proportion of undernourished people in the world.[178] In fact, according to the Baseline Survey of Food Security and Nutrition[179] prepared by the Technical Secretariat for Food Security and Nutrition (SETSAN) in 2004, chronic malnutrition among children under the age of five increased from 36 percent in 1997 to 46 percent in 2006, and 35 percent of households continue to be highly vulnerable to food insecurity.

The above data is corroborated by the Multiple Indicator Cluster Survey (MICS)[180] published in 2009 which reveals that although there has been a significant reduction in chronic malnutrition in the intervening period – from 48 percent in 2003 to 44 percent in 2009, still one out of every two children in Mozambique suffers from malnutrition (41 percent), and 15 percent are underweight at birth.

2. Identifying the Hungry

After the end of the civil war in 1992, Mozambique achieved a drastic reduction in the prevalence of undernourishment. Nevertheless, apart from a series of rapid gains in the initial years, mainly due to the very low starting point, SETSAN reports that the country's food security situation is worsening and that despite strong economic growth and a reduction in poverty, there has not been a parallel reduction in the incidence of chronic malnutrition and that, in fact, more people are hungry. This suggests that market-based development policies alone are not sufficient for human development. SETSAN concluded that social justice, accountability and empowerment principles are missing elements in traditional efforts at poverty reduction.

Over the last few decades, Mozambique has increased its food production and the availability of basic foods, such as maize, cassava and beans. This means that the country requires less food aid and, in fact, due to the gradual move towards economic stability and the growing per capita GDP in the last five years (2005-2009), the number of people living below the poverty line is expected to have diminished from 54 percent to 45 percent.[181] From 2003 to 2009 chronic malnutrition declined from 48 percent to 44 percent; under weight for age reduced from 22 percent to 18 percent; and acute malnutrition (low weight for height) from 5 percent to 4 percent.

---

176 Ministry of Planning and Development. 2006. *Plano de Acção para a Redução da Pobreza Absoluta 2006-2009 (PARPA II)*, Maputo, Mozambique (English version of PARPA II at http://www.imf.org/external/pubs/ft/scr/2007/cr0737.pdf).

177 Mozambique. 2008. *Report on the Millennium Development Goals, p.9.*

178 FAO. 2006. *The State of Food Insecurity in the World 2006*. Rome.

179 SETSAN. 2006. *Baseline Survey of Food Security and Nutrition in Mozambique*. Maputo, Mozambique.

180 Mozambique. 2009. National Institute of Statistics (INE). *Multiple Indicator Cluster Survey (MICS) 2009.*

181 Ministry of Planning and Development. 2006. *Plano de Acção para a Redução da Pobreza Absoluta 2006-2009 (PARPA II)*, Maputo, Mozambique. (English version of PARPA II at http://www.imf.org/external/pubs/ft/scr/2007/cr0737.pdf).

Nutrition and food insecurity indicators, however, although they show positive trends, have not improved as much as the above economic and production statistics would have justified. Low weight at birth is still 15 percent and as much as 41 percent of children suffer from malnutrition.[182]

It is clear that food shortages and hunger are exacerbated by natural disasters such as drought, floods and cyclones. Added to this is the high prevalence of HIV/AIDS, affecting 16 percent of the population[183], and malaria which is responsible for 30 to 40 percent of under-five mortality.[184] An inadequate diet and poor food habits, when combined with high post-harvest loss, can be disastrous, hence the urgent need for a holistic approach to eradicating poverty and hunger.

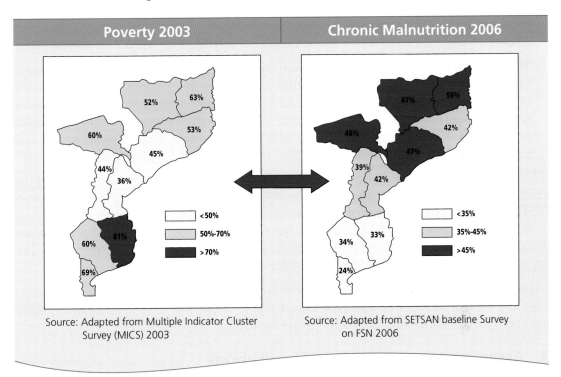

Poverty 2003	Chronic Malnutrition 2006

Legend (Poverty 2003): <50%, 50%-70%, >70%

Legend (Chronic Malnutrition 2006): <35%, 35%-45%, >45%

Source: Adapted from Multiple Indicator Cluster Survey (MICS) 2003

Source: Adapted from SETSAN baseline Survey on FSN 2006

In comparing the two maps above, it is clear that poverty and chronic malnutrition are related but not synonymous. The fact that the highest prevalence of undernourishment is to be found in the northern provinces, where food production is highest and poverty levels (in terms of income) are average, represents the 'Mozambique paradox'. One explanation for this is that people living in the southern provinces of Sofala, Inhambane, Gaza and Maputo have, on average, more financial, social, human, physical and natural capital and are thus in a better position to cope with sudden shocks.

On a very general level, many people are trapped by the 'triple threat' of poverty, food insecurity and HIV/AIDS mentioned earlier and it is difficult, if not impossible, for them to escape this threat.

---

182 Mozambique. 2003. National Institute of Statistics (INE). *Multiple Indicator Cluster Survey (MICS) 2009.*

183 Mozambique. 2004. *Demographic Impact of HIV and AIDS in Mozambique.*

184 Mozambique. 2009. *National Survey of Under Five Mortality.*

In describing the livelihoods of the most vulnerable groups, the 2006 baseline survey indicated the two most disadvantaged of these as being (a) subsistence farmers with few assets, and (b) the informal sector.

(a) **Subsistence farmers with few assets**: People in this category have poor access to all types of resources. There is a high dependency ratio and a high proportion of female headed households (more than 40 percent) or households headed by elderly people (nearly 25 percent). Only one third of heads of households are literate, and about 60 percent have never attended school. For these people, agriculture consists of a few crops only (monoculture). Subsistence farmers with few assets are highly vulnerable and economically marginalized. Six percent of the total rural population fall within this category and are scattered throughout the country although there is a higher concentration in Cabo Delgado, Nampula and Inhambane.

(b) **The informal sector**: The greatest number of households vulnerable to food insecurity belong to the informal sector. They earn most of their income from this sector or from seasonal work on farms, and some manage to produce food for their own consumption. The informal sector is characterized by a low asset base. Geographically, the highest concentration of households belonging to this sector are found in Nampula (20-30 percent), and in Zambezia, Tete and Inhambane (10-20 percent).

Government and other stakeholders made extensive use of the baseline survey in revising the Food Security and Nutrition Strategy (ESAN) in 2006.[185] The survey addressed the link between food production and the incidence of chronic malnutrition, indicating that efforts to increase food production alone have had only a limited effect on hunger reduction. The State needs to improve its services and provide accompanying measures, such as, nutrition education, empowerment, capacity building, hygiene promotion and environmental sanitation.

The baseline survey is considered to be a significant step forward in understanding the poverty issue and the root causes of hunger. As there are many reasons why some people cannot realize their right to food, different policy responses are needed.

3. Assessing Laws, Policies and Institutions

A case study on the right to food in Mozambique was conducted in 2004 for the 32nd session of the Standing Committee on Nutrition[186] and for the evaluation[187] of ESAN I (the first food security and nutrition strategy) in 2005. The study revealed a series of shortcomings that needed to be addressed in a revised strategy. In the earlier strategy food security was perceived as a sectoral rather than a cross-cutting issue. Consequently, ESAN II made an effort to assign clear roles and responsibilities for line ministries. ESAN I, drawn up in 1998, had also lacked a human rights based approach and an institutionalized partnership with civil society groups.

---

185 SETSAN. 2007. *Estratégia e Plano de Acção de Segurança Alimentar e Nutricional 2008-2015 (ESAN II)*. Maputo, Mozambique.

186 Standing Committee on Nutrition. 14-18 March 2005. The SCN four country case studies: *Integrating food and nutrition interventions in national development plans in order to accelerate the achievements of the MDGs in the context of the realization of the human right to adequate food. A synthesis of findings and recommendations.* 32nd Session of the SCN. Brasilia.

187 SETSAN. 2005. *Availação da Implementação da Estratégia de Segurança Alimentar e Nutricional* (Evaluation of Implementation of ESAN). Maputo, Mozambique.

A similar assessment emerged from the first Poverty Reduction Strategy (PARPA I)[188] adopted in 2002. PARPA II, the country's overarching poverty strategy for the years 2006 to 2009 (extended to 2010), differs from its predecessor in the definition of poverty. Whereas PARPA I had a very narrow view of why poverty persisted in Mozambique, had not envisaged the creation of an environment that would enable individuals to rise above the poverty threshold, and did not consider the State capable of delivering services, PARPA II introduced a more holistic approach.

PARPA II is built on three main pillars: *governance, human capital and economic development*. These are much broader than the traditional sectors, such as education, health and agriculture. Together with eight cross-cutting issues, including food security and nutrition, they stress the fact that development policies have to be implemented holistically by adopting a human rights based approach.

PARPA II envisaged the adoption of a right to food perspective, the approval of a Right to Food Law and the inclusion of the right to food in the Constitution by 2007. However, this was considered too ambitious and a more realistic date for the approval of a draft law was set for 2010 without the accompanying prior goal of the inclusion of the right in the Constitution.

In response to the revised goal, SETSAN undertook an assessment of the country's legal framework. Preliminary findings revealed that the right to food is not clearly recognized in the country's legislation, nor is a human rights based approach taken into account in the related laws enacted, with the exception of the Code of Marketing of Breast-milk Substitutes.

The Constitution states in Article 11 (c) and (e) that the realization of human rights and the pursuit of an adequate standard of living for all are State objectives. It also guarantees the right to life (Art. 40), health (Art. 89), consumers' rights (Art. 92), and the right to social protection (Art. 95).

Moreover, the legal assessment shows that the country has ratified most of the binding human rights instruments of importance for the realization of the right to food, such as the Convention on the Elimination of all Forms of Discrimination against Women[189] and the International Convention on the Rights of the Child.[190] It also recognizes the Universal Declaration on Human Rights (Art. 42 of the Constitution). Furthermore, the Government is committed to the Millennium Declaration and adopts the Millennium Development Goals as the overall goals for its development programmes. ESAN II points to the need for Mozambique to ratify the ICESCR, a need that is also echoed in a civil society right to food campaign.

A series of case studies commissioned by FAO in 2004[191] clearly indicated the need for an institution tasked with coordinating the implementation of the right to food. In a sense, SETSAN is performing this function.

SETSAN was created as a follow up to the World Food Summit of 1996, held in Rome. The Summit Declaration stated the need to formulate food and nutrition strategies. The President of Mozambique signed the declaration and decided to immediately embark on this endeavor.

---

188 Ministry of Planning and Development. 2002. *Plano de Acção para a Redução da Pobreza Absoluta 2002-2006 (PARPA I)*. Maputo, Mozambique.

189 Ratified by Resolution n° 4/93, of 2 June 1993. National Assembly, Mozambique.

190 Ratified by Resolution n° 19/90 of 23 October. Council of Ministers, Mozambique.

191 FAO. 2006. *The Right to Food Guidelines: Information Papers and Case Studies*. Rome.

A multisectoral cluster was created in 1997 to prepare the Strategy, and in 1998 Government approved the Food Security and Nutrition Strategy (*Estratégia de Segurança Alimentar e Nutricional* – ESAN), by Decree number 23/98 of 23 December 1998.

As stated in the Strategy, a mechanism was needed to coordinate implementation. Consequently, the Multisectoral Cluster was renamed the Technical Secretariat for Food Security and Nutrition (SETSAN) and became part of the Ministry of Planning and Finance, as of 1998. In 2002, the Secretariat was transferred to the Ministry of Agriculture, and incorporated into the National Directorate for Agrarian Services (DNSA).

SETSAN's major role is that of coordinating food security and nutrition at all levels of governance. It has to ensure the transversality of food security and nutrition, integrating both aspects in all related plans, policies, strategies, programmes and legislation, and also in Mozambique's strategic development. SETSAN is also responsible for promoting (influencing) budget allocations for food security and the right to food, as well as monitoring its implementation, capacity building and awareness raising.

While SETSAN's task is to coordinate and monitor food security and nutrition as well as right to food activities, its convening and coordinating power is seriously limited by its placement within the government. The Secretariat is 'hidden' within a department of the Ministry of Agriculture, which makes it very difficult to promote a cross-cutting topic such as food security, nutrition and the right to food. In addition, food security has been associated primarily with food production – a notion which considerably reduces the motivation of other ministries to get involved. SETSAN has advocated for a more visible and powerful placement for the Secretariat.

Despite this institutional weakness, SETSAN has been carrying out its work along the lines recommended by the Right to Food Guidelines. It covers eleven strategic ministries and autonomous public institutions, such as the Social Action Institute, the National Institute for Calamity Management, the National Council for Combating HIV and AIDS, civil society institutions and academia. SETSAN is co-chaired by the Ministries of Agriculture and Health.

ESAN II highlights the need to rearrange the institutional framework for food security and nutrition, and to reinforce SETSAN's mandate in terms of coordinating food security and nutrition as an autonomous body.

At the time of finalization of this publication, important efforts are underway in view of formally creating SETSAN and strengthening its position as a national food and nutrition security institution with its own juridical and administrative autonomy. It is planned to continue to be mandated with inter-ministerial coordination for the implementation of ESAN and for programmes and activities related to food security, nutrition and the right to food. SETSAN would be chaired by the Minister of Agriculture along with the Minister of Health serving as vice-chair.

4. Sound Food Security Policy

Integration of the right to food in Mozambique's policy framework began with the country's most important policy: the Poverty Reduction Strategy (PARPA II) for the years 2006-2009 (extended to 2010). PARPA II acknowledges food security as a cross-cutting issue and states that every person in the country has a human right to food. The strategy also recalls key human rights principles, such as the principle of equality (including gender equity and non-discrimination), the promotion of participation, transparency and accountability, the dignity inherent in all human beings, the rule of law, and empowerment. It also highlights the need for a holistic approach to food security and nutrition and recognizes health and social protection as human rights.

The definition of the right to food stated in PARPA II, paragraph 210, is as follows:

> 'Everyone has the human right to a standard of living that assures him/her health and well-being. Regular, predictable access to food is a fundamental right of all people and a basic premise for their welfare. Food and nutritional security requires that all people have, at all times, physical and economic access to a sufficient quantity of safe, nutritive foodstuffs that are acceptable within a given cultural context in order to meet their nutritional needs and their food preferences, so that they can lead an active and healthy life. The four components of food and nutritional security are availability, stability of supply, access, and use of the foods.'[192]

The section on monitoring contains the strongest language adopted in the right to food cause. Two food security indicators have been accepted so far: The first, prevalence of underweight children under five years of age, is the accepted MDG 1 indicator.[193] The second is a structural indicator measuring whether or not an adequate legal framework for the right to food would be established by 2007. This indicator boosted the work of Mozambique's right to food promoters: the country's most important strategy not only recognized the right to food, but made an official request to work on the legal dimension of this right.

An evaluation of the implementation of ESAN I carried out in 2006 showed that the strategy drafted in 1998 had serious shortcomings, such as:

* lack of analysis of the links between HIV/AIDS and food security and nutrition;
* absence of clear monitoring indicators, benchmarks or targets included;
* emphasis on the food insecurity problems in rural areas at the expense of urban food insecurity;
* food and nutrition insecurity referred to mainly as an emergency phenomenon and the result of natural disasters;
* little attention paid to the structural vulnerability that is directly associated with the multiple causes of absolute poverty;
* no consistent definition of beneficiaries;
* no operational plan for multisectoral coordination and for the implementation of sector programmes;
* no provision of an implementation budget showing which resource limitations were creating a particular constraint to SETSAN's operations;
* no mechanisms to allow for the strengthening of community involvement and recognition of the concept of the district as the basic unit for planning;
* the heterogeneity of food insecure individuals left without acknowledgment; and
* lack of adherence to a right to food approach.

---

192 Mozambique, *Action Plan for the Reduction of Absolute Poverty 2006-2009* (PARPA II). (Available at http://www.imf.org/external/pubs/ft/scr/2007/cr0737.pdf).

193 MDG1, Target 1.C: 1.8 Prevalence of underweight children under-five years of age; and 1.9 Proportion of population below minimum level of dietary energy consumption (Available at http://www.mdgmonitor.org/goal1.cfm).

As a result of this assessment, SETSAN revised the strategy and the Second Food and Nutrition Security Strategy (ESAN II) was fully accepted by the Council of Ministers in 2007. ESAN II is based on human rights principles[194] and is Mozambique's overarching strategy for food security and nutrition. It differs from ESAN I by adding the notion that food security and nutrition are not to be understood as access to a minimum basket of calories, proteins and other specific nutrients. Rather, they have to do with food safety, the quality, diversity and sustainability of production practices and respect for the local culture. The strategy explicity states that its overall objective is to *'guarantee that all people in Mozambique have, at all times, physical and economic access to an adequate diet that is necessary in order to live an active and healthy life, thereby realizing their human right to adequate food.'*[195]

ESAN II specifies the State's obligations, according to the ICESCR, to respect, protect, and fulfil the right to food. It also adopts, as basic values, the human rights principles of universality, equity, human dignity, participation, transparency, accountability and transversality. Furthermore, the strategy clarifies the roles of the different stakeholders.

The right to food concept calls for an integrated approach involving, on the one hand, the promotion of activities that have the potential to ensure people's access to all the resources needed to ensure food security and nutrition in Mozambique (credit, insurance and social protection, shared natural resources) while, on the other, it protects those who cannot provide for themselves, by creating and strengthening social security networks.

In addition to the common food security pillars (availability, access, stability of supply and utilization), ESAN II emphasizes 'adequacy', looking at what is 'socially, culturally and environmentally acceptable'. The strategy also mentions the challenges it will be likely to face during implementation. Among these are:

- establishing food security and the right to food as central elements of sectoral strategies;
- putting the three levels of State obligations into practice;
- establishing recourse mechanisms;
- implementing ESAN II in a multisectoral and inter-institutional manner;
- basing food security interventions on underlying and root causes of food insecurity; and
- strengthening the role of civil society in monitoring the realization of the right to food.

The Second Food and Nutrition Security Symposium organized by SETSAN[196] in June 2008 approved a declaration that validated the strategy and strongly recommended the allocation of resources for its implementation. Furthermore, government institutions, civil society, donors and the UN agreed to embark on the elaboration of the Right to Food Law, with broad participation throughout the country.

---

194 The title of the strategy reads: *Segurança Alimentar e Nutricional, um Direito para um Moçambique Sem Fome e Saudável.*

195 Mozambique. 2007. *Estratégia de Segurança Alimentar e Nutricional – ESAN II*, Chapter 3.5.

196 SETSAN. 2008. *Declaration of the 2nd Food and Nutrition Security Symposium (June 2008)*. Maputo, Mozambique.

5. Allocating Roles and Responsibilities

ESAN II includes a section on roles and responsibilities. It describes the different types of responsibilities and obligations of State and non-State actors and recalls the obligation of the State to guarantee the human right to adequate food and to establish recourse mechanisms. ESAN II sees an important role for civil society in the field of awareness development and information. It also assigns a role for the private sector and academia in guaranteeing an integrated approach.

Institutions for protecting, monitoring and promoting human rights, such as the Human Rights Commission and the Ombudsman system were established in 2009 and 2007, respectively. The nominations for these positions are still pending in Parliament.

The draft version of ESAN II submitted to the Council of Ministers foresaw the establishment of the National Council for Food and Nutrition Security (*Conselho Nacional de Segurança Alimentar e Nutricional* – CONSAN). This administratively and financially autonomous body would ensure the coordination of all Government activities relevant to food security and monitor whether the different ministries were implementing the activities foreseen by ESAN II.

In 2007, following the positive experience of other countries in locating such a council at supra-ministerial level, SETSAN proposed that the Council be attached to the Prime Minister's office and report directly to Cabinet. In the suggested scheme, SETSAN would then play the role of an executive body, by coordinating the implementation of decisions made. It would be supported by a technical committee composed of line ministry staff and civil society representatives who provide information on the country's food security situation.

The call for a food security and nutrition council was raised again by participants at the Second Food and Nutrition Security Symposium in 2008. The issue is likely to be revisited in the process of formulating the Right to Food Law. SETSAN's long-term strategy is to convert the annual, high-level food security symposium into a permanent CONSEA-type coordination forum.

There have been notable developments at the district level, which is the country's lowest administrative level. District Councils with a civil society majority have been established to examine and approve district development plans. While the capacity needs of these councils are said to be enormous, their very existence is already an important first step. As Council members become better informed, their effectiveness will increase.

6. Integration of Food Security, Nutrition and the Right to Food in the Policy Framework

SETSAN has the mandate to integrate the right to food into the relevant policies and programmes, giving priority to cross-cutting areas referred to in ESAN II. Thus, several policies and programmes, such as the Methodology for the Integration of Food Security and Nutrition in District Development Planning, and the Nutrition Plan of Action, fully endorse the right to food principles. Training in this approach is being provided with support from FAO.

Currently, steps are being taken to integrate the right to food in the National Human Rights Promotion Plan. The inputs for this Plan will be based on what is already included in PARPA II and ESAN II. However, the opportunity will be taken to promote the inclusion of ratification of ICESCR in the Plan, as well as explicit recognition of the right to food in the Constitution.

In addition, the right to food will be included in the Plan of Action for Reduction of Poverty (which replaces PARPA II) in 2010, and in the national Plan of Action for the Reduction of Chronic Malnutrition, also in 2010.

7.   Legal Framework to Realize the Right to Food

The Constitution of Mozambique does not make explicit mention of the right to food and the country has not yet accessed the ICESCR. These facts coupled with weak legal institutions and the absence of laws on the right to food, means that there is only indirect legal protection for this right, at best derived from the constitutional provision on the right to life (Art. 40). The Constitution[197] also recognizes the right to health (Art. 89), consumers' rights (Art. 92), and the right to social protection (Art. 95), as indicated earlier. Human rights protection and socio-economic welfare are also State objectives under Part II, Sections II and III of the Constitution. The Government has approved the National Code of Marketing of Breast-milk Substitutes along the lines of its international equivalent[198] which recognizes the human right to food and children's rights.

Efforts are underway to ensure direct legal protection of the right to food. The PARPA II provision of developing a right to food framework legislation by 2010 has already been mentioned. FAO is supporting SETSAN, the Ministry of Planning and Development, the Ministry of Justice and the civil society network ROSA (*Rede das organicações da soberania alimentar*) in this endeavour. A taskforce, coordinated by SETSAN which includes high-level representation from the relevant ministries, has been set up to draft the law. The process is highly participatory and includes training and awareness raising to build consensus and facilitate approval of the law. Work is based on the Guide on Legislating for the Right to Food, published by FAO's Right to Food Unit in 2009.

The right to food is a new concept adopted in PARPA II but, despite some awareness raising and capacity building conducted by SETSAN and FAO, it is still not thoroughly understood by many stakeholders. Consequently, the taskforce working on the Right to Food Law is aware of the need to improve understanding of the right to food concept in general and the framework law in particular. This will also facilitate the formulation, approval and implementation of the law.

Experience from other legislative and policy-making processes have shown that the inclusion of institutions at all levels of governance, together with the public in general, will ensure better results and a smooth iterative process. Thus, one of the steps was to hold a seminar with the ministries concerned (Agriculture, Health, Justice, Planning and Development, Women and Social Action, Education, Culture, State Administration, Industry and Commerce, Environmental Coordination, Fisheries, Public Works, Transport and Communications), UN agencies in the field and civil society, to make them aware of the process and to solicit their inputs.

The participatory process for developing the right to food law also includes extensive consultations at district level. These include capacity building and information dissemination to both duty bearers and right holders. Full participation of public officials at district and community levels is

---

197 Mozambique. 1990. *Constituição da Republica*. Maputo, Mozambique (Portuguese and English version available at http://www.mozambique.mz/pdf/constituicao.pdf and http://confinder.richmond.edu/admin/docs/moz.pdf).

198 World Health Organization. 1981. *International Code of Marketing of Breast-milk Substitutes*. Geneva.

crucial for the implementation of the future law. These are the levels where both duty bearers and right holders meet and where the implementation of the right to food will make a difference in the lives of the people concerned.

The preliminary findings of the process to develop the Right to Food Law already suggest that **Parliament** is one of the most important stakeholders. While the Executive Branch plays a major role in the actual drafting of the new law, Parliamentarians are those who have to finally discuss and approve it. Involving them right from the start and equipping them with knowledge and explanations as to what right to food is, what purpose it serves and the need for it, increases the likelihood of its rapid approval after submission. Interaction with them at the beginning of the process already revealed some of the major challenges that need to be addressed.

Seminars will be held with parliamentarians to discuss the contents of the law, advocate for its approval, and familiarize them with the implementation of the right to food in other countries. Media actors and the private sector are also involved in the process. Contributions by the latter include disseminating messages on food security and nutrition and the legislative process in the context of their corporate communications and through various media – such as radio, television, newspaper and magazine spots – thus combining publicity with information dissemination.

The Second Food and Nutrition Security Symposium in 2008 proposed embarking on a framework law rather than a specific law, in recognition of the cross-cutting character of the right to food and the need to mainstream its norms and principles in all of the country's legislation. This implies that after the approval of the right to food law, Mozambique has to conduct a compatibility analysis of its sectoral laws.

The right to food framework legislation will influence present and future legislation in areas such as land, water and natural resources, health, nutrition and consumers' rights. SETSAN and civil society groups are already lobbying for the inclusion of right to food provisions in current draft bills, such as the proposals regarding the Statute of the National Commission on Human Rights, the Law on the Rights of People Living with HIV/AIDS and the Consumers' Rights Act.

Information and training are the prerequisites for making key actors aware of the right to food. A lesson drawn from laws recently approved by Parliament (such as the Act on the Protection Against Trafficking of Human Beings)[199] shows that training and advocacy facilitated a clear understanding of the implications of the Act, an informed discussion and a smooth passage through Parliament. Legal advocacy seems to have an effect on other fronts as well. Following a two-year intensive campaign by civil society actors to achieve the ratification of the ICESCR, the Ministries of Foreign Affairs and Justice are now intensifying the debate on this issue. The inclusion of the right to food in the Constitution is also recommended by civil society.

8. Monitoring the Right to Food

Some legal provisions are included in the monitoring matrices of PARPA II and ESAN II. While the former has already been mentioned, the latter refers to recourse mechanisms and inclusion of the right to food in the Constitution.

---

199 Mozambique. *Act no. 6/2008*, of 9 July 2008.

In order to improve the process by which PARPA II is formulated and implemented, Mozambique sought stronger participation from civil society. A solution was found in the creation of the 'Observatorio da Pobreza' or Poverty Observatory. This civil society platform enhances citizen participation and social accountability processes in implementing and monitoring PARPA II.

The Poverty Observatory was envisioned as a consultative platform for dialogue on poverty reduction, implementation of PARPA II, and improved governance. It is intended to feed information to the national government and parliament through SETSAN. The Observatory has been operational since 2002 and comprises representatives of the national government, international donors and local civil society. The G20, whose name emerged from the 20 CSOs that participated in the first Poverty Observatory in 2003, today comprises over 400 organizations and networks. It conducts a complementary poverty analysis (Relatorio Annual da Pobreza) annually.

SETSAN's food security monitoring function does not yet follow a human rights based approach. Its working group on vulnerability assessment and monitoring publishes a vulnerability assessment four times a year. This is a technical document describing the broad trends at provincial level.

As a result of the Second Food and Nutrition Security Symposium held in Maputo in June 2008, academics are seeking to develop a Right to Food Observatory to be created at the Centre for Human Rights, Faculty of Law, Eduardo Mondlane University. This institution will undertake research assessments and will promote civic education as well as legal and policy contributions on the right to food.

9. Legal and Administrative Recourse Mechanisms

During the process of elaboration of ESAN II, there was a discussion on the establishment of recourse mechanisms, however, consensus was not forthcoming. This issue, which is part of the workplan to implement ESAN II, was postponed for further debate during the drafting of the law.

The creation of the Human Rights Commission[200], in keeping with the Paris Principles, is certainly a milestone in the promotion and protection of the right to food in Mozambique. The institution will be assigned powers to receive cases of human rights violations, report on the matter to the international human rights bodies and decide on the mandatory involvement of government institutions. The Commission members will be elected by the National Assembly. Another such initiative was the creation, in 2007, of the Ombudsman's Office[201] under Parliament, which will receive cases and recommend action to be undertaken by government institutions. The Ombudsman will also be elected by the National Assembly.

10. Capacity Building

As in many other countries, civil society pressure in Mozambique initiated the right to food movement. The topic was pushed mainly by the Human Rights League (Liga dos Direitos Humanos). Other NGOs in the Civil Society Network, ROSA, included the right to food in their advocacy work. However, their campaign was relatively weak, given the country's under-funded and donor-dependent NGO environment.

---

200 Mozambique. Act no. 33/2010, of 22 December 2009.

201 Mozambique. Act no. 7/2006, of 16 August 2007.

SETSAN has already incorporated the right to food in its daily work. During the preparation of materials for training on food security and nutrition at provincial level, a module on the right to food was included for local administrators and the media. Training will be extended to all 128 administrators as well as to key professionals in the media.

The strategy of the SETSAN-based Right to Food Project is to advocate for and promote a human rights based approach to food security and nutrition in a progressive manner. Simultaneously, the project prepares awareness-raising material and events for the general public, and offers strategic and systematic capacity building to the food sovereignty network, ROSA, and other national and locally-based NGOs. The media is included as a stakeholder in this regard. In 2007, two training sessions on the right to food were organized for journalists in Maputo and Beira, and publications on food security have improved in quality as a result. Additional training and sensibilization sessions for the media are being planned.

SETSAN will also adopt solution-oriented learning in the process of elaborating the Right to Food Law. All participants will be challenged to contribute and provide concrete inputs regarding issues included in the law. The FAO Guide on Legislating for the Right to Food will be used for building the capacity of civil servants and civil society members involved in the formulation process.

Academia has also embarked on promoting the right to food. The Right to Food Project, in partnership with the Centre for Human Rights of the Faculty of Law at the Eduardo Mondlane University (UEM), organized a seminar for 26 lecturers from 12 universities throughout the country, and some civil society activists. This led to the adoption of the right to food as the topic for the traditional Moot Court Competition (a fictitious human rights violation judgement) amongst law students from these universities in 2009, including a seminar on the right to food and the rights of the child.

Universities have integrated the right to food as a topic in the teaching of fundamental rights. In addition, the Master's course in the Faculty of Law at the UEM, together with the Master's course in Human Rights in the Faculty of Law at the University of Pretoria have integrated the right to food in their curricula, inspired by the *Right to Food Curriculum Outline* published by FAO in 2009[202]. Both professors and students will support SETSAN in drafting the right to food framework law.

11. Conclusions

After the approval of PARPA II, the greatest success in the implementation of the right to food in Mozambique has been the approval of ESAN II in 2007 and its subsequent launching in 2008. It adopted a human rights based approach and pursued the objectives of the international human rights instruments, with a special focus on ICESCR, General Comment 12 and the Right to Food Guidelines.

The government has acknowledged the weaknesses in the legal framework, thus, corrective measures are being undertaken. What makes a difference in the implementation of the right to food in Mozambique is the fact that this concept and food security and nutrition are seen to be complementary to one another. For example, PARPA II adopts a comprehensive approach towards

202 http://www.fao.org/righttofood/publi_en.htm

the right to food. It highlights the regular and predictable access to food as a fundamental right and basis for welfare. It also adopts a definition of food and nutrition security as the right to have – at all times – physical and economic access to a sufficient quantity of safe, nutritive foodstuffs acceptable within the cultural context and nutritionally appropriate based on the needs and preferences of each individual in order to live an active and healthy life. Hence, the concept of right to food, food in general and nutrition to be specific, are interlinked in such a way as to give the right to food the implication of the existence of four elements of food security – availability, access, utilization and stability. Aside from PARPA II, ESAN II complements the definition of right to food and how it is implemented by linking HIV/AIDS and food security and nutrition, climate change, gender and women's rights. It takes into consideration rural vulnerability; moreover, it calls for the strengthening of institutional capacity and an increase in resources allocated to food security and nutrition. It also sets up adequate evaluation, accountability and monitoring procedures.

As far as explicit acknowledgment of right to food is concerned, Mozambique is planning to have a draft Right to Food Law by the end of 2010. The task force working on this piece of legislation includes government, civil society, academics and other stakeholders. The process has been both consultative and participatory. Furthermore, a right to food framework legislation is among the priorities of SETSAN and the Ministry of Justice. An amendment to the Constitution and accession to ICESCR are also being considered by the government. The establishment of the National Commission on Human Rights will greatly facilitate the implementation of this right.

The greatest challenge will be to use this enabling policy and legal framework for the benefit of right holders, especially the vulnerable, and to ensure increased budgetary allocations for the right to food. SETSAN is striving to strengthen its provincial offices and is considering deploying district representatives. This will be performed in line with a central government that puts strong emphasis on sub-national planning and implementation. However, due attention has to be paid to training district officials and civil society representatives on the right to food and food security.

Another enormous challenge relates to right holders' capacity to claim their rights and to duty bearers' ability to respond to their obligations. Despite the progress made, there is still a lack of understanding among stakeholders regarding the right to food and its practical implications. Capacity building and dissemination of information will continue to be of paramount importance.

Finally, justiciability of the right to food is another step being taken into account. With the creation of the national human rights institutions referred to above, the country has introduced the initial administrative and quasi-judicial claim mechanisms for human rights. Judicial mechanisms are also in place, such as the General Attorney Cabinet, the Judicial Court, the Administrative Court and the Labour Court. These mechanisms do not explicitly refer to the right to food; but it is assumed that a lawyer can address claims in all three courts depending on the character of the violator and the 'competence' of the case.

The government's achievements in improving the policy, institutional and legal framework have been considerable. However, this work is not yet complete. The main challenge will be to use these frameworks for the benefit of the people and help them to feed themselves and realize their right to food.

Recommendations

- A food security and nutrition strategy based on the right to food must be truly comprehensive in its approach. The definition of food security established at the 1996 World Food Summit incorporates the complementary concepts of food security, nutrition and the right to food through four pillars of food security – access, availability, utilization and stability.
- Right to food as a strategy must be defined by taking into consideration the concept of food security, but also by incorporating concerns related to nutrition, climate change, gender and women's rights, all of which have an influence on access, availability, utilization and stability of the entitlement to food.
- Rural populations are often considerably more vulnerable and should not be overlooked in creating right to food policies.
- Institutional structure, capacity, and resources for the purpose of promoting food security and nutrition are key in effective and adequate evaluation, accountability and monitoring efforts.
- Outreach and capacity development for civil society as well as for parliamentarians are essential to increase the likelihood of a right to food law to be highlighted and approved.
- The media is an essential tool for disseminating information on new law and people's rights. Full advantage should be taken of this means of communication.
- The right to food is a new concept for many countries. Legislative drafting processes require investment in capacity development and awareness raising. Therefore, the process is a long-term undertaking. Possible delays should not be seen strictly as a disadvantage; rather, they are a necessary demand of time that eventually leads to readiness to adopt the new law and an increase in the likelihood of its efficient and sustainable implementation.

VI. UGANDA
Joining Forces for the Right to Food

Main Points

- The prevalence of poverty in Uganda declined from 56 to 31 percent from 1992 to 2006[203], yet the actual number of those suffering from undernourishment increased from 3.6 million to 4.4 million during the same timespan.[204] The country's continued vulnerability and food insecurity is mainly due to armed conflict, demographic changes and poverty-related issues.
- Following ratification of the International Covenant on Economic, Social, and Cultural Rights (ICESCR) in 1987, Uganda recognized the right to adequate food in its Constitution adopted in 1995 and in its progressive Uganda Food and Nutrition Policy (UFNP) adopted in 2003. The UNFP expressly recognizes and pledges to realize the right to food.
- The Uganda Food and Nutrition Strategy and Investment Plan (UFNSIP) was subsequently developed in 2005 and efforts are currently underway to adopt the Food and Nutrition Act – a framework law to support legally-binding obligations to realize the right to food. This law will also enhance the establishment of the Uganda Food and Nutrition Council (UFNC), a national multi-sectoral agency designed to coordinate the implementation of the UFNP.
- Through the implementation of the Poverty Eradication Action Plan (PEAP), Uganda is moving towards ensuring food security for all and is thereby strongly supported by civil society groups, NGOs and the country's Human Rights Commission.

1. Background

Uganda actively participated in the work of the Inter-Governmental Working Group for the development of the Right to Food Guidelines. Like other member countries of the FAO, it pledged to apply the Guidelines in its efforts to realize the right to food in the context of national food security.

FAO has supported activities in Uganda to promote awareness of the right to food and implement the Guidelines. Some of these initiatives also centered on the development of methodological tools for monitoring the right to food at country level. With the aim of promoting the utilization of the Guidelines at all levels, FAO facilitated the preparation of a case study for developing a manual linking financial allocation with right to food entitled 'How Budget Analysis Can Strengthen Right to Food Advocacy'. FAO has provided support towards the drafting of the principal right to food legislation known as the Food and Nutrition Bill which, if adopted, will become Uganda's Food and Nutrition Act. Civil society seminars organized by the FoodFirst Information and Action

---

203 Uganda Bureau Of Statistics, *2010 Statistical Abstract*.
   (Available at http://www.ubos.org/onlinefiles/uploads/ubos/pdf%20documents/2010StatAbstract.pdf).
204 FAO. 2009. *The State of Food Insecurity in the World*. Rome.

Network (FIAN) have been held to promote general awareness of the Guidelines and to prepare a monitoring report on the status of implementation of Uganda's right to food commitments.

2. Identifying the Hungry in Uganda

The Uganda Bureau of Statistics (UBOS), in coordination with the Ministry of Agriculture, Animal Industry and Fisheries (MAAIF) and the Ministry of Health (MOH), is mandated to assess the country's food and nutrition security situation. Food balance sheets also provide valuable information on levels of calorie (energy) consumption per capita per day, and are thus essential in planning right to food relevant programmes.

Uganda has seen a progressive reduction in poverty in recent years. In 1992, more than half of the population (56 percent) were poor. That number has declined by 25 percentage points. According to UBOS, Uganda had a poverty headcount of 31 percent in the year of 2006[205] and projections hover around 28 percent for 2010. A slightly different measure of poverty as presented by the UNDP Human Development Report shows a Human Poverty Index of 34.7.[206] Although Uganda's performance on poverty reduction is significant, there exists a geographic difference in poverty incidence within the country. Whereas 13.7 percent of the urban population is poor, 34.2 percent are suffering from poverty in rural areas according to the State of Uganda Population Report of 2007.[207] Furthermore, in spite of reasonable economic growth in the last decade, inequality seems to have increased at a national level.

According to the World Bank, Uganda's GDP growth rate has gone from 7.3 percent in 1992 to 10.8 percent in 2006 and remains projected at 8 percent for 2010.[208] On the other hand, inequality as measured by the Gini coefficient has increased from 0.37 in 1992 to approximately 0.41 in 2006.[209] This tendency towards a large gap between incomes earned by the rich and those earned by the poor is more pronounced in urban areas where the Gini coefficient as of 2009 was measured at 0.43 versus 0.36 in rural areas.[210]

Cross-border food exports between Uganda and conflict-affected areas of neighbouring countries, such as the Democratic Republic of Congo, Kenya, Rwanda and southern Sudan, have fuelled rising domestic food prices with further negative consequences on the capacity of the poor and vulnerable to access adequate food. In addition, although strong monetary policy has led to a decrease in the inflation rate from 42 percent in 1992 to 12.6 percent in 2009, Uganda's population currently suffers low purchasing power with the annual inflation rate well above the Bank of Uganda target rate of 5 percent.

---

205 Uganda Bureau Of Statistics, *2010 Statistical Abstract*.
   (Available at http://www.ubos.org/onlinefiles/uploads/ubos/pdf%20documents/2010StatAbstract.pdf).

206 http://hdr.undp.org/en/media/HDR_20072008_EN_Indicator_tables.pdf

207 Uganda Population Report of 2007 referenced in the National Development Plan of the Republic of Uganda, Table 6.6, p. 185 (Available at http://www.imf.org/external/pubs/ft/scr/2010/cr10141.pdf).

208 World Bank – World Development Indicators on Uganda.
   (Available at http://data.worldbank.org/indicator/NY.GDP.MKTP.KD.ZG).

209 Republic of Uganda. 2010. National Development Plan (2010/11-2014/15), Table 2.2, p. 16.
   (Available at http://www.imf.org/external/pubs/ft/scr/2010/cr10141.pdf).

210 *Ibidem*, Table 6.6, Pg. 185 and Table 2.2, p. 16.

Over a span of two decades from 1979 to 1999, the percentage of undernourished people in Uganda declined from 33 to 19 percent; however, between 1990-92 and 2004-06, the actual number of undernourished people rose another 800 000 from 3.6 million to 4.4 million.[211] During the same period dietary energy supply (DES)[212] increased from 2 270 to 2 370 kilocalories per person per day by 2004-2006,[213] pointing to the fact that food supply available for human consumption was indeed sufficient for the people of Uganda but that access to food remained an issue. This growth in food availability coupled with sustained economic growth rates as mentioned above, and an increase in the number of undernourished people, concurred with a population increase of 3.4 percent on average. Although there may be a need to control population growth, there must also be efforts aimed at studying the structural root causes of hunger in order to increase access to food for the undernourished. Furthermore, income-generation schemes, investment in education and creation of employment opportunities to fight food insecurity in the long term are also imperative.

An assessment of the food and nutrition status in Uganda was undertaken jointly by MAAIF and MOH in 2004. This assessment provided data for the formulation of the Uganda Food and Nutrition Strategy and Implementation Plan (UFNSIP)[214] formulated in 2005. In addition, poverty-related nutrition vulnerability information was reported in the context of a poverty status analysis undertaken in 2003 and 2005 by the Ministry of Finance, Planning and Economic Development (MFPED). Vulnerability was categorized based on three causal factors: armed conflict, demographic and poverty-related concerns.[215] Food insecurity in relation to lack of access to food is widespread among populations affected by internal and cross border armed conflicts.

Prior to 2009, there were over one million internally displaced persons (IDPs) in the north and north-east regions of Uganda and over 200 000 refugees relying on humanitarian food aid.[216] Food insecurity is also a common phenomenon among orphans and families living with HIV/AIDS – a pandemic affecting over 2 million (6.8 percent) people.[217] Efforts are under way by the Ministry and the UBOS to undertake a second such survey.

In the north-eastern semi-arid region of Karamoja, periodic droughts have persisted over the past years threatening to cause famine: human and livestock mortality are reportedly increasing, due to the scarcity of food and water. In addition to drought, the proliferation of small fire arms (guns) within regions where nomadic lifestyles and movement of livestock are common has resulted in brutal tribal warfare over the years. Households have been forcefully deprived of access to their

---

211 http://www.fao.org/fileadmin/templates/ess/documents/food_security_statistics/country_profiles/eng/Uganda_E.pdf

212 DES refers to the total food available for human consumption in a country. It is expressed in kilocalories per capita per day. See also FAO. 2008. *The State of Food Insecurity in the World*. Rome.

213 FAO. 2009. *The State of Food Insecurity in the World*. Technical Annex. Table 2, p. 53. Rome.

214 Republic of Uganda, *The National Food and Nutrition Strategy, Final Draft*, November 2005. (Available at http://www.health.go.ug).

215 This category was initially reported in the 2003 Poverty Status Report by the MFPED. It was also adopted in the draft Uganda Food and Nutrition Strategy and Investment Plan of 2005.

216 FAO. 2004. *Right to Food Case Study: Uganda*. IGWG RTFG/INF 4/APP.4: Table 5, p.16. Rome. (Available at http://www.internal-displacement.org/countries/uganda). Since then, two-thirds of the 1.5 million IDPs who lived in camps have returned to their place of origin.

217 MOH and ORC Macro. 2005. *National HIV/AIDS Sero Behavioral Survey*. (Available at http://www.measuredhs.com/pubs/pdf/AIS2/AIS2.pdf).

own food resources, due to cattle rustling, armed robbery, and periodic tribal clashes over pasture lands making Karamoja one of the poorest and most insecure regions in the country.

The figures below show lack of significant improvements in malnutrition indicators. In Uganda, stunting, an indicator of chronic undernutrition, still affects almost 40 percent of children under the age of five.

Malnutrition among children aged between 6 and 59 months (%)

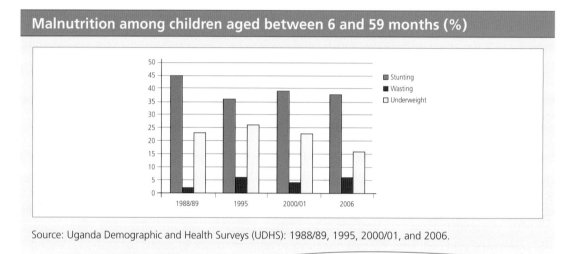

Source: Uganda Demographic and Health Surveys (UDHS): 1988/89, 1995, 2000/01, and 2006.

Although vulnerable groups have been identified, there are no disaggregated food security data available on specific vulnerable groups, such as children, the elderly, the chronically ill, ethnic minorities, persons with disabilities, and others. This problem is related to the fact that Uganda lacks a comprehensive national data system. There is a current initiative to improve the quality and reliability of food and agriculture statistics in the country in order to harmonize national, sub-national and meta level data with the aim to combat hunger. The initiative known as Country STAT[218] was launched by FAO and is being implemented by UBOS in conjunction with MAAIF.

> *"Systematic and scientific data collection and management systems, especially national identification data, are needed to inform important decisions and duty bearers on how to intervene in favour of all vulnerable groups, if the human right to adequate food is to be realized …"*

Joel Aliro Omara, former Commissioner, Uganda Human Rights Commission

What is apparent from the Ugandan experience is the need to undertake a specific right to food baseline assessment, with indicators covering both the processes and outcomes affecting the realization of this right. This type of assessment should include the impact of existing food and nutrition security, social security, poverty eradication, and people's empowerment programmes. Furthermore, already available dietary assessment methods may offer relatively simple tools for identifying food insecure and vulnerable groups.

3. Assessing Laws, Institutions and Policies

According to Right to Food Guideline 17.2, States may consider conducting right to food impact assessments in order to identify the impact of domestic policies, programmes and projects on the progressive realization of the right to adequate food for the population at large and for vulnerable groups in particular. This has not yet taken place in Uganda at the State level. However, independent assessments have been conducted by non-state actors as a means of promoting awareness of the right.[219]

Evidence points to the emergence of an enabling political, social, and economic environment in Uganda, including measures to restore democratic governance, peace and human rights, through the establishment of an independent human rights commission as well as an active international cooperation framework with bilateral and multilateral donors and non-governmental organizations. In addition, significant steps forward with regard to policy and legislative reforms have focused on food security, nutrition, and the right to food. The Uganda Food and Nutrition Policy (UFNP) that expressly recognizes the right to food was adopted in 2003, and a Food and Nutrition Security Strategy and Investment Plan (UFNSIP) was drafted in 2005 as a multi-sectoral coordinating mechanism to further the objectives of the UFNP. However, implementation of the Policy remains weak in the absence of enactment of the Food and Nutrition Bill which is the legal basis for the Policy.

The absence of minimum wage laws has contributed to an increase in the number of low-paid workers in both the formal and informal sectors, and among the urban poor and landless urban squatters. This may seriously constrain vulnerable population groups from gaining access to the variety of foods needed in order to have a nutritionally adequate diet.

The overriding challenge in the Ugandan experience has been a gap between policy statements and practice. Overarching macro-economic growth policies, especially poverty reduction strategies, have not taken into consideration human rights principles and approaches in their implementation.[220] Furthermore, Uganda has not submitted any report on the status of economic, social and cultural rights (ESCR) as required under Article 16 and 17 of the International Covenant on Economic, Social, and Cultural Rights (ICESCR). The 5th report was due in June 2010[221].

This may point to insufficient institutional capacity to undertake an ESCR assessment, thus reinforcing the need to strengthen advocacy and capacity building programmes. In this context, the Right to Food Guidelines can serve as a capacity strengthening tool.

A similar example of lack of institutional capacity is Uganda's lack of implementation of the 2003 African Union Summit 'Maputo Declaration on Agriculture and Food Security' which commits

---

219 For example, occasional papers on the impact of policies and legal frameworks on Ugandans' right to food were presented at a national Right to Food Seminar organized by the Uganda Human Rights Commission, the MAAIF and Makerere University. The results of the national seminar served as an input in the preparation of a case study on Uganda's right to food situation. In addition to providing significant information to policy makers and implementers, the Uganda case study report (February 2004) also provided information to the Intergovernmental Working Group on the Right to Food Guidelines.

220 See the Right to Food Case study on Uganda presented to the Intergovernmental Working Group that developed the *Right to Food Voluntary Guidelines* (Available at http://www.fao.org/righttofood/kc/downloads/vl/docs/AH258.pdf).

221 See details on Uganda's reporting status on ICESCR (Available at http://www.unhchr.ch/tbs/doc.nsf/NewhvVAllSPRBy Country?OpenView&Start=1&Count=250&Expand=182.5#182.5).

African countries to allocate not less than 10 percent of national budgets to agriculture and food security. In fact, agriculture is among the least-funded sectors, having continuously received less than 5 percent of the country's national budget.[222]

4. Sound Food Security Strategies

Uganda's Poverty Eradication Strategies and the Right to Food Guidelines

In 1997, Uganda, like other Highly Indebted Poor Countries, developed a Poverty Reduction Strategic Paper (PRSP) which was subsequently adopted as the Poverty Eradication Action Plan (PEAP). The PEAP constituted Uganda's overall national development and institutional policy and planning framework to eradicate poverty through economic growth, as the overriding objective of all national development programmes. The second poverty reduction strategy paper, PEAP 2004/5 – 2007/8, set the target of reducing income poverty to 10 percent by 2017, and was built on five pillars:

- Economic management
- Enhancing production, competitiveness, and incomes
- Security, conflict resolution, and disaster management
- Good governance, and
- Human development

The policy options and actions under each pillar were very relevant to the realization of the right to food in Uganda, and potentially could have had both positive and negative effects. The main thrust of **the first pillar** was boosting economic growth through maintaining macroeconomic stability and promoting the private sector. Implementation of economic development policies should have given consideration to the Right to Food Guidelines particularly Guidelines 2.4, 2.6 and 2.7. These Guidelines suggest a comprehensive approach in economic policy making which ideally must provide for adequate social safety nets, methods to improve livelihoods, efforts to increase institutional capacities as well as guarantees for market functioning and regulatory frameworks. The Guidelines promote investment in rural areas where poverty and hunger are predominant with the aim to improve access to factors of production such as land, capital, and technology. This should occur along with the promotion of bottom-up participation in economic policy decisions whilst responding to the rising trends of urban poverty.

The second pillar of the PEAP 2004/5 – 2007/8 promoted and value added agriculture and socio-economic transformation as a means to eradicate poverty. Under this pillar, the Plan for Modernization of Agriculture (PMA) was established to enhance productivity, competitiveness and incomes within the PEAP framework. The aim of the PMA is to improve performance in seven priority areas: agricultural research; agricultural advisory services; agricultural education; access to rural finance; marketing and agro processing; sustainable natural resource use and management; and rural infrastructure. It does not target the economically deprived, vulnerable poor but aims to empower the economically active poor who have land, or the capacity to engage in commercial agriculture. The PMA also aims to achieve food security through the market (commercialization)

---

222 Republic of Uganda, MAAIF, CAADP Brochure 5 – October 2009 (Available at http://www.nepad-caadp.net/pdf/ stocktaking%20-%20uganda.pdf). Also see FAO report on progress towards the Maputo Declaration (Available at http://www.fao.org/docrep/meeting/007/J1604e.htm).

without taking into consideration national social safety nets to protect the right to food of the ordinary subsistence farmers, who have limited or no capacity to produce for the market. Such methods of support for vulnerable groups and creation of safety nets are outlined in Guidelines 13 and 14.

The third pillar focused on IDPs in northern Uganda and improving their food and nutrition security status. The realization of peace in this area of the country is very important for the food security of more than one million IDPs who have not been able to engage in meaningful agriculture and commerce for more than two decades, relying on humanitarian food aid alone. The existence of IDPs is a sensitive political issue but it is important to ensure that IDPs have access at all times to adequate food according to Guideline 16.5. Guidelines 16.1 and 16.2 also state that food should be considered a national strategic resource but not one to be used as a bargaining tool to create political and economic pressure. Referring to the 1949 Geneva Convention and its 1977 Additional Protocols, the Right to Food Guidelines assert the importance of humanitarian law and assurance that starvation not be used as a political tool. In 1999 the Cabinet of Uganda approved a National Disaster Preparedness Policy and Institutional Framework which was revised in 2003. Subsequently, in consultation with the Uganda Human Rights Commission (UHRC) and UN agencies, the government adopted a National Policy on Internal Displacement and a Disaster Preparedness and Management Policy Framework which were to be implemented through the Office of the Prime Minister. They were intended to provide a coordinated framework in response to emergencies that require humanitarian and relief food aid. The IDP Policy committed the State to protect the rights and entitlements of IDPs during displacement and promoted their resettlement and re-integration into society. The Policy placed the burden of responsibility of protecting the food security of IDPs on the State, however, it has not been effectively implemented with reference to the majority of IDPs that were uprooted as a result of conflict between the State and the Lord's Resistance Army. Furthermore, as of 2009, although more than half of IDPs returned to their original territories, this massive return has outpaced recovery planning in Uganda.[223] As for the UN Peace Building and Recovery and Development Program (PRDP) for Northern Uganda and its objective to align UN intervention with Uganda's policy framework on IDPs, the results are yet to be evaluated.

The fourth pillar on good governance aims to pursue democratization, to ensure respect for human rights, to create government accountability and transparency, and to provide an effective and efficient judicial system. Good governance is explicitly recommended in various Right to Food Guidelines, in particular, through Guideline 1 entitled "Democracy, good governance, human rights and the rule of law" and Guideline 7 on legal frameworks as well as Guideline 18 on encouraging States to use national human rights institutions to further the realization of the right to food. Existing institutional structures in Uganda have not assumed responsibilities specific to the right to food due to budgetary and capacity constraints. The UHRC has not prioritized right to food monitoring due to inadequate resources, while the Justice, Law and Order Sector reform process, the Law Reform Commission, and the Parliament of Uganda have not taken significant action to monitor the right to food provisions of international human rights instruments to which Uganda is a State Party.

---

223 http://www.internal-displacement.org/8025708F004CE90B/(httpCountries)/04678346A648C087802570A7004B97 19?OpenDocument

The fifth and last pillar emphasized the importance of human development through improvements in education and skills development, maternal and child healthcare, water and sanitation services, social development, and cross-cutting issues – particularly malnutrition, HIV/AIDS, and population growth. These goals parallel those recommended by the Right to Food Guidelines such as Guideline 11 on education, Guideline 8 on access to resources and assets, and Guideline 10 on nutrition. Uganda developed a National Health Policy in 1999 providing the basis for developing Health Sector Strategic Plans (HSSP I & II) aimed at the improving childhood nutrition and access to primary health care. However, nutrition care for other vulnerable groups has not been considered under the HSSP. Uganda has implemented the Universal Primary Education programme which aims at ensuring human development through education for all.

Policies Supportive of the Right to Food

After more than ten years of negotiations and debate amongst multi-sectoral actors represented on the Uganda Food and Nutrition Council (UFNC), the Uganda Food and Nutrition Policy (UFNP) was adopted in July 2003 under the MAAIF and MOH.[224] The UFNP expressly recognizes the right to adequate food and pledges to support its progressive realization in the country.[225]

In the foreword section of the UFNP, the MAAIF and MOH reiterate Uganda's commitment to address food and nutrition insecurity by stating: *'Government is committed to fulfilling the Constitutional obligation of ensuring food and nutrition security for all Ugandans. This Food and Nutrition Policy is therefore important as it provides the framework for addressing food and nutrition issues in the country.'* [226]

In the 'Background' section, the Policy acknowledges international human rights instruments that recognize the right to food, specifically mentioning Article 25(1) of the Universal Declaration of Human Rights[227] and Article 11(1) and Article 11(2) of the ICESCR.[228] The UFNP further elaborates on the genesis of the right to food debate in international conferences and the World Food Summits of 1996 and 2002.

As Uganda is a party to the ICESCR (ratified in 1987), the UFNP pledges to guide the implementation process of the right to food. Statutory Objective XXII of the 1995 Constitution also deals with the commitment to achieve optimal food and nutrition security in order to attain the highest achievable standard of health and wellbeing for all people in Uganda.[229] The provisions concerning the implementation of the UFNP are underpinned by human rights principles and provide the

---

224 http://www.fao.org/righttofood/inaction/countrylist/Uganda/Uganda_foodandnutritionpolicy.pdf

225 A three-day national seminar was held at Nile Resort Hotel, Jinja district, from 22 to 24 January 2003 on the theme *'Towards the Implementation of the Right to Adequate Food in Uganda'*. It drew stakeholders from Government institutions, international and national civil society organizations, the private sector, and farmers. It was jointly organized by the UHRC, MAAIF, Makerere University, and the Oslo based International Project on the Right to Food in Development. The discussions from the Jinja seminar contributed significantly to the final text on the right to food in the UFNP. (Available at http://www.ajfand.net/IssueIV%20files/IssueIV-News%20Bits%20-%20National%20seminar.htm).

226 Republic of Uganda. 2003. *The Uganda Food and Nutrition Policy.*
    (Available at http://www.fao.org/righttofood/inaction/countrylist/Uganda/Uganda_foodandnutritionpolicy.pdf).

227 http://www.un.org/en/documents/udhr/index.shtml

228 http://www2.ohchr.org/english/law/cescr.htm

229 http://www.ugandaembassy.com/Constitution_of_Uganda.pdf

responsible actors with the necessary mandate to advance human rights measures related to combating hunger.

UFNP's objective is to improve the nutritional status of all through coordinated multi-sectoral interventions focused on food security and nutrition, together with increased incomes: the overall goal is to achieve a healthy nation and sustainable social and economic wellbeing. The Policy indicates twelve focal areas for intervention: (1) Food supply and access; (2) Food processing and preservation; (3) Food storage, marketing, and distribution; (4) External food trade; (5) Food aid; (6) Food standards and quality control; (7) Nutrition; (8) Health; (9) Information, education and communication; (10) Gender, food and nutrition; (11) Food, nutrition and surveillance; and (12) Research.

The absence of a formally and legally established Uganda Food and Nutrition Council (UFNC) and Secretariat has hindered the implementation of the UFNP at the level of local government. The Food and Nutrition Bill currently under consideration does provide the basis for the establishment of the necessary institutional framework and would potentially provide the key towards implementation of policy initiatives at the local level. The UFNP foresees coordination between local councils and the national council regarding the establishment of food and nutrition committees at district and lower levels, and the generation of food security and nutrition data for use in the planning of national policies.

Other policies supportive of the right to adequate food in Uganda include: the 1999 National Health Policy, implemented by the MOH; the National Orphans and other Vulnerable Children Policy of 2004, implemented by the Ministry of Gender, Labour, and Social Development; the National Policy for Internally Displaced Persons of 2004, implemented by the Office of the Prime Minister; and the Uganda National Culture Policy 2006. A draft National Land Policy is under review by stakeholders as of 2009, however, it has generated divisive debate and opposition from protective cultural institutions and members of the public who see the process as a non transparent means to 'grab' land from the vulnerable poor.

Uganda Food and Nutrition Strategy and Investment Plan and the Right to Food

The UFNSIP was developed by the UFNC, and endorsed by MAAIF and MOH in November 2005. It is currently awaiting approval by Cabinet before it can be tabled in Parliament for debate. The UFNSIP focuses on: advocacy for good governance; inter-sectoral coordination; empowerment of duty bearers and right holders; policy decentralization; and gender equality – targeting to provide nutrition support to all women of child-bearing age. Once fully adopted, the UFNSIP will in effect provide a platform for the government to commit political, financial, and administrative resources to fulfil its legally binding international and national obligations related to food security and nutrition, including the realization of the right to food for all communities and households throughout the country.

Having recognized the right to adequate food and re-affirmed Uganda's commitment to realizing this right, the proposed UFNSIP elaborates why and how food and nutrition insecurity are issues of policy concern. It discusses the causes and lifecycle consequences of childhood malnutrition and identifies the country's nutritionally vulnerable groups based on three causal factors –

vulnerability due to demographic or health status, to conflict, and to livelihood strategies. A number of important elements are still missing: (i) the identification of monitoring indicators; (ii) the financial capacity and other relevant resources needed to effectively implement the UFNP; (iii) an investment plan; and (iv) the necessary tools and resources to ensure a well-developed implementation strategy. It is imperative to speed up the approval of the UFNSIP so as to provide the platform and agenda for the realization of the right to food.

5. Allocating Roles and Responsibilities

Uganda's political commitment to progressively realize the right to food can be traced back to 1987 when the country ratified the ICESCR. The UFNC was established that same year, albeit on an *ad hoc* basis, to advise the government on food and nutrition policy planning, development, programming, implementation, research, and monitoring and evaluation. Thirteen institutions/ sectors are represented in the Council to enable effective representation and participation. The chair is appointed by the MAAIF. However, the Council lacks the necessary operational structures and legal mandate to assume these responsibilities at present. The draft Food and Nutrition Bill when enacted should establish the profile of Council members and recommend that the UHRC has a seat on the Council.

Institutions such as the UHRC have been established for the promotion of good governance principles and respect for human rights. The Commission is mandated under Article 52 (10) (h) of the 1995 Constitution as an autonomous institution for monitoring State compliance with international human right instruments and treaties to which Uganda is a party. The Justice, Law and Order Sector is a reform process within the judicial sector that has been established to ensure the humane treatment of all prisoners and their access to justice through coordination between the various institutions under the judicial arm of the State. Moreover, there is a well-established, autonomous 'Inspectorate of Government', managed by an 'Inspector General of Government', with a mandate to fight corruption through investigation and prosecution as well as the naming and shaming of individuals or institutions found to be corrupt or otherwise engaged in corrupt activities.

Uganda in 1997, as part of its second poverty reduction plan, created a PMA to transform the country's agriculture from subsistence to commercially oriented sector. The PMA provides a strategic framework and guidelines for both sectoral and inter-sectoral policy and investment planning at both local and national levels and supports the MAAIF in ensuring a stable food and agricultural economy. The Uganda National Bureau of Standards (UNBS) collaborates with the MOH in ensuring food safety, disease prevention and control, adequate care and, ultimately, improved nutrition security. Institutional objectives and mandates specific to the right to food are not emphasized within Uganda's institutional framework; rather, they are considered as milestones to be achieved gradually through food and nutrition security policy objectives.

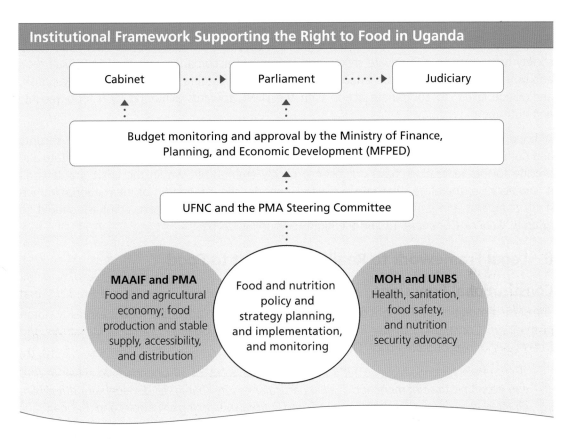

Institutional Framework Supporting the Right to Food in Uganda

Cabinet ∙∙∙∙∙∙▶ Parliament ∙∙∙∙∙∙▶ Judiciary

Budget monitoring and approval by the Ministry of Finance, Planning, and Economic Development (MFPED)

UFNC and the PMA Steering Committee

MAAIF and PMA Food and agricultural economy; food production and stable supply, accessibility, and distribution

Food and nutrition policy and strategy planning, and implementation, and monitoring

MOH and UNBS Health, sanitation, food safety, and nutrition security advocacy

The PMA Secretariat and Steering Committee continue to play a significant role in facilitating policy, strategy, legislative and planning processes within the *ad hoc* UFNC framework. This may provide an opportunity for prioritizing budgetary interventions for the right to adequate food. Despite its supportive role, however, the PMA has been criticized for emphasizing a predominantly commercial agenda that includes empowerment of farmers already organized and those considered active poor only, and achieving food security through commercial agricultural development without considering the vulnerable poor who cannot afford to produce for the market.[230]

This analysis re-affirms the need for the UFNC and PMA to prioritize strategies and activities aimed at prioritizing domestic food production, and the empowerment of right holders to access and claim adequate food, with specific emphasis on rural areas. Since hunger is predominant in the north and north-eastern districts of Uganda due to conflict and drought, existing institutional frameworks need to develop action plans that ensure an equitable distribution of food supplies between food surplus and food deficit areas.

---

230 The agriculture sector as a whole grew by 0.7 percent in 2006/2007 while the food crop sub-sector declined by 0.9 percent. However, it was projected to rise, due to increased budgetary commitments and the restructuring of the National Agricultural Advisory Services, so it can target food crops production as a means of improving household food security. Refer to: Ministry of Finance, Planning, and Economic Development (MFPED). 2008. *Background to the Budget 2008/2009*, p. 13 (See http://www.finance.go.ug).

Within the UFNC framework, private-public partnerships are involved through emphasis on effective representation of all food and nutrition stakeholders, including representatives from civil society organizations, the private sector, and farmers. In accordance with the UFNP, local food and nutrition councils are to be established at district, constituency, county, sub-county, parish, and village levels, to coordinate action with the UFNC towards achieving local food security and nutrition.

Although the sector roles and responsibilities of the UFNC institutions focus on food security and nutrition objectives, the right to food obligations to respect, protect, and fulfil (facilitate and provide for) its realization, have not been explicitly emphasized within the UFNC framework. It is imperative, therefore, that relevant actors consider the integration of international human rights standards and principles into activities of the UFNC. In this context, emphasis should be given to General Comment 12 and the Right to Food Guidelines.

6. Legal Framework to Realize the Right to Food

Constitutional Provisions

The State recognition of the right of every person in Uganda to adequate food is expressly stated in the 1995 Constitution of the Republic of Uganda under Uganda's *National Objectives and Directive Principles of State Policy*. Objective XIV states that:

> *'The State shall endeavour to fulfil the fundament rights of all Ugandans to social justice, and economic development and shall in particular ensure that (a) all development efforts are directed at ensuring the maximum social and cultural well-being of people; and (b) Ugandans will enjoy rights and opportunities and access to education, health services, clean and safe water, work, decent shelter, adequate clothing, food security, pension and retirement.'*

In addition, Objective XXII of the Constitution specifically deals with food security and nutrition emphasizing that:

> *'The State shall (a) take appropriate steps to encourage people to grow and store adequate food; (b) establish national food reserves; and (c) encourage and promote proper nutrition through mass education and other appropriate means in order to build a healthy State'.*

Although food security concerns are apparent, the actual right to adequate food is not expressly stated as a fundamental right in Chapter 4 of the Constitution entitled *Protection and Promotion of Fundamental and Other Human Rights and Freedoms*. Thus, it is interpreted to be lacking in adequate legislative support. It is rather a constitutionally mandated statutory objective implied in Article XIV and Article XXII as well as in the protection implied within the body of fundamental rights incorporated within the Constitution. Thus it can only be realized gradually through policy implementation. Therefore, Uganda's legal framework is made up of isolated legislative acts with no remedial and recourse mechanisms in the case of right to food violations or violations of other human rights.

National Legislation Relevant to the Right to Food

The following list of legislation has been identified as relevant to the progressive realization of the right to food:

- The *Food and Drug Act* (1964): essentially regulates food quality and safety.[231] However, the National Drugs Authority (NDA) transformed the *Drugs* component into the *Drugs Act* of 1993, while the MOH is currently spearheading an effort to draft a food safety bill.
- The *Uganda National Bureau of Standards Act* (1993): establishes the Bureau as an institution that monitors and regulates food quality and safety, based on international *Codex Alimentarius* standards and guidelines.
- The *Water Statute* (1995, reviewed in 1997): aims at ensuring a clean, safe and sufficient supply of water for domestic use by all Ugandans.
- The *Children Act, Chapter 59 of the Laws of Uganda* and the *Penal Code Act*, Sections 157 and 158: place obligations on the State and on parents to ensure adequate care, including the provision of food for all children in their care, and make it an offence for parents to neglect their child-care responsibilities.
- The *Uganda Bureau of Statistics Act No 12* (1998): provides for the assessment, analysis, and dissemination of all national statistics obtained through surveys;
- *Adulteration of Produce Act* (2000): makes it criminal to adulterate food.
- *The Plant Protection Act* (2000): ensures the use of environmentally friendly farm chemicals in the care of plants and animals.
- *The National Agricultural Advisory Services Act* (2001): provides for the institutionalization of NAADS as an implementation pillar of the PEAP within the PMA.
- *The Land Act* (2004) (Amendment): clarifies four types of land ownership systems – customary, communal, *Mailo* and freehold land ownership systems, and provides for compulsory land acquisition of un-utilized and underdeveloped land by the Government for development purposes.
- *Agricultural Chemicals Act* (2007): provides regulations for the use of agriculture chemicals so as to protect agricultural bio-diversity and thus facilitate sustainable and viable food systems.

The Food and Nutrition Bill, 2009

The Government of Uganda originally drafted the Food and Nutrition Bill in 2003 which was revised in 2008. The current draft of 2009 awaits adoption.[232] It was developed as a step towards ensuring that everyone is free from hunger and will progressively enjoy the right to adequate food. Once adopted, the Food and Nutrition Act will also be used as a major instrument for implementation of the national strategy outlined in the UFNSIP.

The objectives of the Bill are:

(a) to recognize, promote, protect and fulfil the right to food as a fundamental human right;
(b) to provide a legal basis for implementing the Uganda Food and Nutrition Policy;

231 For details on the Food and Drugs Act refer to ftp://ftp.fao.org/codex/ccafrica16/ca1606ae.pdf. Uganda's laws can also be accessed through the National Online Law Library (Available at http://www.ugandaonlinelawlibrary.com/lawlib/law_index.asp).

232 Food and Nutrition Bill. 2009 (Current draft available at http://www.health.go.ug/nutrition/docs/population/FOOD_AND_NUTRITION_BILL_2009.pdf).

(c) to plan, budget and implement the Uganda Food and Nutrition Policy using a rights based approach and to ensure the participation of rights holders and the accountability of duty bearers;

(d) to ensure that food is treated as a national strategic resource;

(e) to promote the policies on food and nutrition as part and parcel of the overall national development policy;

(f) to ensure the integration of the needs of the vulnerable in food and nutrition strategies;

(g) to promote public education and sensitization on food and nutrition especially in rural areas to enhance the impact on food and nutrition security; and

(h) to promote the drawing up of strategies to respond to food and nutrition concerns at all levels of government.

Recognizing that adequate food is a fundamental human right, with entitlements to be accorded to everyone without discrimination, the Bill prohibits starvation of Ugandan citizens, and ensures that all Ugandans should be free from hunger and under-nutrition. It also ensures that vulnerable groups – such as the elderly, infants, children, orphans, refugees, internally displaced persons, pregnant and nursing mothers, people with disabilities, sick persons with chronic diseases such as HIV/AIDS, victims of conflict, prisoners, rural people in precarious livelihood situations, marginalized populations in urban areas – are accorded basic nutrition by the State. Equality and protection against discrimination are also aspects clearly promoted and emphasized in this piece of legislation as it requires the State to perform its obligations towards vulnerable groups on an equal basis as compared to the rest of its citizenry.

The emphasis on anti-discrimination along with other directives for State officials can also be seen in the guiding principles of the Bill that are set to govern the implementation of the Bill once it is enacted into law. They are:

(a) equity and non discriminatory physical and economic access to food;

(b) coordinated efforts of public authorities and their full participation in related food and nutrition security issues;

(c) accountability of duty bearers and transparency in the food sector particularly emergency food aid, through open access by the public to timely and reliable information on decisions and actions taken;

(d) people participation in the planning, design, implementation and monitoring and evaluation of decisions that affect them; and

(e) the use of scientific research and evidence based decision making.

The Bill emphasizes the role of public authorities and provides a standard of scrutiny which the State is required to meet in respecting, protecting, promoting and fulfilling the right to food in the country. It is this criteria that sets a benchmark within which the State should work, firstly by directly providing food when the most vulnerable individuals are not able to have access to it for reasons beyond their control, and secondly, by taking concrete steps to ensure that the right to food is progressively realized in the country.

The Bill outlines measures to ensure that food is available and accessible to all; that food is not only safe but also complies with required nutritional value; that education and information services about proper nutrition are available and accessible as well for the entire population;

and that the issue of who should be responsible for school feeding is also addressed. As a general rule, it states that parents are responsible for providing school meals to pupils in primary schools. It requires heads of households to ensure that all members of their homesteads have enough food reserves and engage in gainful work to prevent famine and poverty. It places the burden of feeding the family on heads of households whereas the State is deemed responsible for directly providing food to those particularly vulnerable who are unable to access food on their own.

This Bill will be the tool to legally formalize the Uganda Food and Nutrition Council as a national multi-sectoral agency to coordinate the implementation of the right to food. It is established with the objective to provide for food security and adequate nutrition for all people in Uganda and to ensure their health as well as social and economic well-being.

In a nutshell, the Bill:

- Gives legal recognition of access to adequate food as a fundamental human right, to which everyone is entitled without discrimination;
- Grants reasonable rights which can be legally enforceable by those whose rights are not met;
- Underscores the legal obligation of the State to meet the food needs of those who lack capacity to access food for reasons beyond their control;
- Creates a clear legal duty of the State to take steps to ensure that the right to food is protected in the country.

The Bill provides a legal mandate to implement the UFNP and it is a welcome effort to address the need to consolidate efforts to combat food and nutrition insecurity in the country. It is also a major tool through which the realization of the right to food should be significantly advanced.[233]

7. Monitoring the Right to Food

Despite its mandate to monitor human rights under Chapter 4 Section 51 of the 1995 Constitution, the UHRC is constrained by inadequate resources. The Commission has been continuously receiving limited budgetary allocations (in some cases less than 50 percent of its budgeted finances) from the central government.[234] In addition, its capacity might be limited by the fact that the right to food is not enumerated as a specific right within the Constitution. The Commission is overstretched while at the same time it is not sufficiently decentralized, that is, it has offices in only 10 out of the 94 districts of Uganda. Consequently, it may have difficulty in playing a role in advancing human rights monitoring – especially with respect to right to food – as an institutional priority in the prevailing budgetary, institutional, and socio-political circumstances.

Violations and abuse of people's right to food in Uganda have been reported in various contexts. In its 2005 report to Parliament, the Commission noted that underfeeding – meaning providing one meal a day or none at all – was reported to be common practice in government prisons. This is a clear violation of prisoners' right to food and indeed a breach of domestic and international law protecting prisoners' rights. In its 2008 report, the Commission acknowledged an improvement in the provision of food to prisoners in government prisons.

---

233 FAO. 2009. *The Right to Food in Uganda – How to make it happen*. Newspaper Supplement. Rome.
234 See UHRC reports (Available at http://www.uhrc.ug).

Spearheaded by FIAN, civil society organizations, in association with the UHRC, made significant contributions to capacity building on right to food monitoring through national workshops held in 2007. FAO supported the UHRC in the development of a right to food assessment tool that can be adapted at the district level to assess the capacity of authorities in identifying and responding to right to food violations[235]. This tool contains process, impact, and outcome indicators that can be taken into consideration in a sound monitoring process at district and community level and which can also be used to identify and map vulnerable groups.

8. Legal and Administrative Recourse Mechanisms

In order for the right to food to become a reality, those whose rights have been violated must have access to remedial measures, including proceedings before a court or other institution that provide restitution, compensation, satisfaction, or guarantee of non-repetition. In Uganda there is no jurisprudence for the human right to adequate food, which would otherwise compel the State to provide adequate food to the most vulnerable poor and those whose right to food has been violated.

Several clauses in the draft Food and Nutrition Bill emphasize the role of public authorities with regard to respecting, protecting and fulfilling the right to food. Such clauses detail provisions of legal redress and administrative review, including cases in which people are aggrieved by the actions of public authorities. Persons aggrieved by decisions taken by the UFNC shall appeal to the Prime Minister or the Minister responsible for agriculture.

With reference to children's right to adequate food, under Chapter 59 of the Laws of Uganda, known as the Children Act,[236] there exists a duty to maintain a child based on which parents and guardians of children are expected to provide for an adequate diet among other necessities. In addition, there is a duty to report violation of children's rights for citizens who have knowledge of a violation.[237] However, the country suffers from lack of resources to implement the provisions of the Act. Uganda does not have the institutions or financial resources to equip authorities to handle children in custody and customary law conflicts with the Act to the detriment of children.

This is not to say that cases are denied. In fact, there have been cases adjudicated under criminal courts in favour of children's rights. For example, criminal case No. CR1376 of August 2000, heard in the Buganda Road Chief Magistrate's Court in Kampala, as well as its criminal appeals Case No. 78/2000 in the High Court, provides insightful jurisprudence on how a 12-year old child was deprived of his right to food and rewarded retribution by a court of law. Although the two defendants appealed to the High Court against the conviction, the trial judge rejected the appeal but reduced the sentences.

Uganda's judicial system comprises a hierarchy of judicial and administrative redress mechanisms. Courts that deal specifically with family and child issues and cases regarding access to land may provide a strong platform for seeking right to food jurisprudence in Uganda. Judicial institutions at the bottom of the pyramid provide greater access to justice for the right to food and other

---

235 http://www.fao.org/righttofood/publi10/UGANDA_assessment_EN.pdf

236 *Children Act* 1997, Chapter 59 (Available at http://www.ulii.org/ug/legis/consol_act/ca19975995).

237 *Ibidem*, Part II, Sections 5 and 11 (Available at http://www.ulii.org/ug/legis/consol_act/ca19975995).

related ESCR, ensuring remedy at local council and family courts. However, right claimants' access to these courts is constrained by high legal fees and inadequate legal representation in the rural areas where most of the courts are located. An additional constraint is the absence of national legal frameworks through which to compel the State to support rural poor people's access to justice.

A Legal Aid Project has been introduced in Uganda to support vulnerable poor people whose right to access justice is constrained by the cost of legal fees. This project is a Uganda Law Society undertaking, with support from the Norwegian Bar Association.[238] However, it operates at higher-level courts which are frequently located in urban centres and focuses on providing legal aid in cases relating to violation of civil and political rights. A mechanism is required whereby right to food violations and abuses can also be protected within the Legal Aid Project framework.

9. Capacity Building

Despite a functional human rights commission and institutions of higher education for the training of lawyers and human rights advocates, the level of civic education on ESCR is still low. This can be attributed mainly to Uganda's turbulent political history characterized by military coups that prevented the enjoyment of civil liberties. As a result, civic educators have emphasized civil and political rights at the expense of ESCR, as a means to provide an enabling political environment for the realization of all human rights and freedoms.

In developing the required capacity, conceptual and ideological barriers associated with the right to food will need to be overcome. In order to ensure freedom from hunger and starvation, institutional capacity building efforts will be necessary, targeted at duty bearers and right holders alike. Capacity development should be undertaken in a transparent, participatory and accountable manner and supported by existing or expanded legal, policy and institutional frameworks.

10. Conclusions

By targeting food insecurity and poverty through the implementation of national policies and initiatives focused on human rights, Uganda may progressively realize the right to food by ensuring that the poor are successfully empowered to raise themselves out of poverty and claim their rights. Despite positive initiatives aimed at achieving socio-economic transformation, the originators of the PEAP have not yet recognized the right to food nor advocated for a human rights based approach for PEAP implementation.

Within the framework of poverty eradication, the National Agricultural Advisory Services may well make a significant contribution to sustainable food production by empowering farmers with improved production and farming methods. However, this will be feasible only with the necessary capacity and tools to mainstream human rights, especially the right to food, within its operations.

Furthermore, the MFPED will play a key role in approving financing for the Uganda Food and Nutrition Council, its staffing and operations. Once established, the Council will serve as the apex for guidance and coordination of all food and nutrition activities in the country and will advise the government in all matters pertaining to food and nutrition.

---

238 For further information see Uganda Law Society online reference page about the Legal Aid Project. (Available at http://www.uls.or.ug/legalaid.asp).

A third and more recent poverty eradiation plan known as the National Development Plan of 2010 remains to be implemented but it goes without saying that the Right to Food Guidelines provide the necessary tools for a human rights based approach to all national efforts in policy making. The integration of the Guidelines into national schemes and the enactment of the Food and Nutrition Bill will be of particular importance in ensuring that the State complies with its human rights obligations and that vulnerable groups become agents of their own development.

A more recent poverty eradiation plan known as the National Development Plan of 2010 remains to be implemented. It goes without saying that the Right to Food Guidelines provide the necessary tools for a human rights based approach to all national efforts in policy making. The integration of the Guidelines into national schemes and the enactment of the Food and Nutrition Bill will be of particular importance in ensuring that the State complies with its human rights obligations and that vulnerable groups become agents of their own development. In particular, the Food and Nutrition Bill will constitute a major milestone in Uganda because it will clarify a number of issues, such as institutional concerns, that are a prerequisite for the implementation of policies and strategies.

Recommendations

- Providing a legal basis for implementing a comprehensive food and nutrition policy incorporating right to food concerns at all levels of government requires the adoption of laws that not only recognize and protect the right to food but address institutional concerns. These concerns can range from the existence of a legal basis for planning, budgeting and implementing right to food policies and programmes – as is seen in the case of the Food and Nutrition Bill of Uganda – and formulating directives on how to promote right to food within the overall national development policy of a country.
- Financing advisory bodies – such as the Uganda Food and Nutrition Council – and equipping them with resources for staffing and smooth operations should be considered a priority. Once established, such entities can serve as the apex for guidance and coordination of all food and nutrition activities in the country and create policy coherence in all matters pertaining to food and nutrition.
- Capacity development is crucial for the realization and monitoring of the right to food. Establishing institutions that support democratic governance and the rule of law, such as the Uganda Human Rights Commission and the Law Reform Commission, is appropriate and necessary to strengthen right to food platforms.
- One recommended path to create capacity is the development of targeted learning modules and curricula based on the Right to Food Guidelines. Right holders, especially those belonging to vulnerable groups, must be informed about their rights and ways to claim them.
- A specific right to food baseline assessment would allow for better targeting of programmes and better monitoring of food insecurity and of human rights violations. Vulnerability assessments should include dietary indicators and be complemented by human rights indicators.
- Development of national data banks containing relevant disaggregated data on the population is a key information tool and must be financed as a prerequisite to designing appropriate policies and programmes.
- Apart from recognition of the right to food, the issue of access to food and to resources to produce or procure food must be addressed as a prerequisite for the realization of such a right.

VII. CONCLUSIONS

In taking an overview of case studies presented in this publication, including the remarks made on the occasion of the Right to Food Forum's topical sessions, what is obviously telling is the underlying theme of human rights principles without which right to food strategies and policies cannot successfully overcome the challenge of fighting hunger in the new millennium. Of the PANTHER principles, the most evident ones with respect to the right to food are participation and empowerment. Whereas in most development circles the word participation would tend to refer to civil society's general participation in decision making processes, in the context of right to food the targeted population of most vulnerable and food insecure people are highlighted – but not at the exclusion of the rest of society – rather simply to remind policymakers and development experts that the goal is to reach those who are normally not even in the realm of societal recognition as people who can contribute to development. These are people who do not have a voice in how a country governs their livelihoods. The hungry need to have more to say about what this situation means for them and what they really need in order to fight hunger. This is stressed in the case of Brazil reaching out to the rural population, granting them citizenship rights, and allowing them the space to claim their right to food. Just as indigenous people in Brazil are now finally recognized as a segment of the population that needs specific attention and should not be ignored, so too Dalits in India have now entered the discussion as potential beneficiaries of policies guided by right to food objectives.

But in order to make such segments of a population – the hungry – become part of the solution of hunger, a tremendous amount of advocacy, awareness raising and educational efforts is required. Such efforts can empower not just the hungry *to speak on their own behalf* in decision making, in designing, in implementing and monitoring policies for the right to food. They can also strengthen the capacity of legislators, policymakers, parliamentarians, and increase their understanding on how the right to food as a conceptual framework and as a legal entitlement can enhance governmental efforts to fight hunger at all levels – local, state and national. All cases presented, especially Mozambique and Uganda, show the relevance of education, training and awareness raising. Knowledge is the first step in empowerment and empowerment is what leads to change within a society towards better food security governance and towards achieving food security for all.

As a result of empowerment, accountability enters the discussion as the key link between declaring a commitment to rights and implementing actual policies on right to food in a transparent and non-discriminatory fashion. In Brazil, for example, municipal councils which comprise community members work to monitor public officials at the local level. In India the accountability mechanisms are in place largely through a series of orders by a Supreme Court that circumvents the power structure underlying the political question of hunger. India serves as an example of a country where legal action can be used to promote right to food and the Right to Food Campaign is an example of an informed civil society that is empowered enough to set the agenda on right to food.

Charity must be replaced with empowerment both in terms of knowledge and in terms of providing the economic and social opportunities for the poor to lift themselves up from a vulnerable situation.

It is, therefore, important to note that economic and social policies are equally relevant in creating coherence and support for a national commitment to a right to food or any food security strategy as a matter of fact. Human development outcomes cannot be achieved without coherent national policies addressing hunger through a variety of sectoral policies, all of which potentially affect the access, availability, utilization and stability of food – four elements identified as key for food security outcomes. A typical example is that of Brazil creating a demand for its own supply of food by linking farmers to the school feeding program. Another example is India where the employment policy links the poor and jobless to public works projects.

Aside from using economic and social policy to promote the right to food, public servants in charge of implementing policies must be capable of coordinating the work they do across all sectors and ministries. This means, a national engine of leadership such as a President declaring his or her support of a country free from hunger, cannot be enough. The workings of government and its institutional structure are key. Finding the right path with which to integrate the right to food into all sectoral policies requires advisory and coordinating bodies on the right to food to be placed in the right location as high up within a hierarchical structure of government. Intergovernmental coordinating bodies are of primary importance but have no power unless they are clearly assigned a level of autonomy, have a legal mandate and receive sufficient budgetary allocations to carry out their mission. The case of Guatemala's SESAN institution and the SINASAN law come to mind as an example of this hurdle that must be overcome to create the enabling environment for right to food.

Aside from coordination of government entities, the legal framework of a society must support the rule of law – not only through jurisprudence on right to food, but also through working institutions such as human rights commissions supported by the national leadership. Such institutions must have a level of autonomy that guarantees human rights are respected throughout implementation and delivery of services without the unnecessary mismanagement and corruption. Legislative processes require a tremendous amount of time but the time invested in adopting right to food laws and creating the institutions that enforce them is well invested towards a key human rights principle of accountability. Non-formal and formal processes both play a role in making rights a reality – the ability to claim them, and the ability to enforce the laws that are drawn to protect them. In this sense, right to food as a development policy option is a guarantee that food is not simply available in the market but reaches the table of the poorest of the poor.

Right to food clarifies the rights, obligations and mandates of all stakeholders allowing victims of violations to exercise power and seek remedy. It allows civil society to scrutinize government and participate in decision making as well as implementation of policies. It promotes government officials to take responsibility and gain trust through transparency. It ensures that the institutional structure both within the legal system as well as within the overall structure of government supports the realization of the right to food through coherence and cooperation. It also expands our understanding of hunger beyond the limited scope of neoclassical macroeconomics. The latter is evident in Uganda as we look at its first and second poverty reduction strategy papers and follow the development of a policy move towards right to food through the establishment of the Uganda Food and Nutrition Council. There is, often, a gap between political commitment for right to food reflected in policy and putting that commitment into practice through institutional, legal, and financial measures. This question is raised in Uganda where the Council does not have effective power unless the Food and Nutrition Bill is adopted. Thus, we can conclude that human rights principles and the right to food fill that gap by putting governments to the test and by addressing the obstacles along the path towards making food security a reality for all, while at the same time providing an overall framework to achieve the vision of a world that is free from hunger and malnutrition.

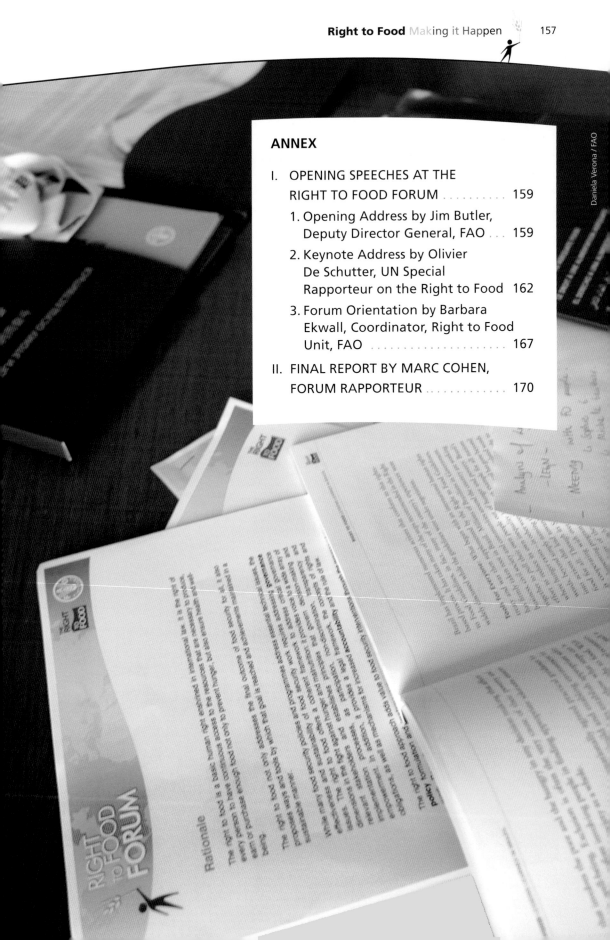

ANNEX

I. OPENING SPEECHES AT THE RIGHT TO FOOD FORUM

Here below are the texts of the three principal speeches delivered at the opening of the Right to Food Forum, which set the scene for the work on hand. The Opening Address was delivered by Jim Butler, Deputy Director General, FAO; the Keynote Address by the United Nations Special Rapporteur on the Right to Food, Olivier De Schutter; and the Forum Orientation by Barbara Ekwall, Coordinator, Right to Food Unit (now Right to Food Team), FAO. Many of the other important contributions are summarized in the Synthesis of the Panel sessions (Part TWO of this report).

1. Opening Address by Jim Butler, Deputy Director General, FAO

Excellencies, Ladies and Gentlemen,

On behalf of the Director-General of FAO, Jacques Diouf, I welcome you warmly to Rome, to FAO and to the Right to Food Forum. The presence of so many participants from all over the world reflects your commitment towards the right to food and the importance of this issue, especially in the context of the present food security crisis.

Message 1: *FAO is committed to the right to food*.
Eleanor Roosevelt, when elaborating the human rights catalogue that subsequently shaped the UN Human Rights Declaration adopted by world leaders almost 60 years ago, said: "Human rights is not something that somebody gives to you, it is something that nobody can take from you."

The right to food is a human right. It is the right of every man, woman and child to be able to produce or procure safe, nutritious and culturally acceptable food, not only to be free from hunger but also to ensure health and wellbeing. It is not charity, nor is it the right to free hand-outs.

Freedom from hunger is one of the fundamental goals set out in FAO's Constitution. At the World Food Summit in 1996, Heads of State and government reaffirmed '*the right of everyone to have access to safe and nutritious food, consistent with the right to adequate food and the fundamental right of everyone to be free from hunger.*' They also committed to the full implementation and progressive realization of this right in order to ensure food security for all.

It was in the follow-up to this commitment that, in 2004, the FAO Council unanimously adopted the Voluntary Guidelines to Support the Progressive Realization of the Right to Adequate Food in the Context of National Food Security (Right to Food Guidelines).

Message 2: *The right to food approach emphasizes good governance*. The Guidelines are a practical tool reflecting international consensus about what needs to be done in some nineteen different policy areas to progressively realize the right to food. They provide a coherent set of recommendations that aim at creating an enabling environment so that everyone can feed himself or herself in dignity. They also define modalities for providing food to those who are unable, for reasons beyond their control, to feed themselves. By looking at rights, institutions and human rights principles, the Guidelines attempt to tackle the root causes of hunger.

The effectiveness and sustainability of food security work requires that governance issues be addressed. The right to food offers a coherent framework to address these critical governance dimensions in the fight against hunger and malnutrition. It provides a voice to the marginalized and to a wide array of relevant stakeholders. It establishes the principles that govern decision-making and implementation processes, such as participation, non-discrimination, transparency and empowerment. Finally, it provides a legal framework, the concepts of rights and obligations, as well as mechanisms for increased accountability and the rule of law.

FAO activities in this area have focused on information and capacity development, the development of methodologies and implementation tools, policy advice and expertise, as well as mainstreaming the right to food into FAO's work. As this work progressed, initial experience was gained in a number of countries, through the implementation of concrete measures to implement the right to food. The objective of the present Forum is to exchange these experiences and lessons learned, and discuss ways to strengthen future implementation of the right to food. The Forum shall demonstrate, with practical examples, how the right to food can contribute to promoting food security for all.

The right to food underpins food security work. It adds value to food security interventions by focusing on issues of voice, participation and accountability in the process of policy formulation and implementation. As reflected in the Guidelines, it builds on the four pillars of food security – availability, access, stability of supply and utilization – with human rights perspectives.

Regarding the process, the approach contributes to strengthening relevant public institutions, integrates partners such as civil society organizations, human rights commissions, parliamentarians and government sectors not dealing with agriculture, and provides further justification for investment in hunger reduction. It contributes to creating and maintaining political will. Promoting the right to food means enhancing government action by introducing administrative, quasi-judicial and judicial mechanisms to provide effective remedies, by clarifying the rights and obligations of right holders and duty bearers and by strengthening the mandate of the relevant institutions. Furthermore, it means strengthening the coordination of food security initiatives and increased policy coherence.

Implementation of the right to food requires a solid partnership between governments, civil society organizations, the private sector, and other relevant stakeholders. This is reflected in the participation at this Forum of representatives from the different sectors involved in right to food work.

Message 3: *Applying the Right to Food Guidelines will improve the response to the present food crisis.* The debate that you will be holding is particularly important in the present climate of high food prices and increasing food insecurity in the world.

The rise in world food prices has, in recent months, pushed the issues of hunger and food insecurity to the top of the international agenda. Soaring food prices have led to a global food crisis, with strong negative social and economic impacts – especially in low income and least developed countries. Poor people typically spend between 50 and 80 percent of their income on purchasing food and will be affected disproportionately by the increase in food prices. A recent study prepared by FAO shows that women are particularly affected.

The Right to Food Guidelines provides recommendations for countries to both understand the food insecurity situation and to shape the response to the present crisis. Through right to food assessments and monitoring, governments can identify the populations at risk.

Appropriate policies, strategies and legislation can be formulated to focus on food insecurity and strengthen the governance of food systems. Institutional capacities and coordination mechanisms, combined with participation and empowerment, make it possible to obtain large buy-in by all relevant stakeholders, policy coherence, and timely and efficient government action.

The High-Level Conference on World Food Security organized by FAO in June 2008 recognized the link between the right to food and the food security challenges the planet is facing at present. It also recognized the importance of urgent international response and cooperation to help developing countries deal with the impact of high food prices. In the outcome document of this Summit, the Right to Food Guidelines are reaffirmed as a framework for the policy response and for measures taken to meet these challenges.

FAO estimates that rising prices have plunged an additional 75 million men, women and children below the hunger threshold, bringing the estimated number of undernourished people worldwide to 923 million in 2007. An enormous and resolute global effort, as well as concrete actions to tackle the root causes of hunger, will be required in order to reduce the number of hungry people and achieve the MDGs.

2. Keynote Address by Olivier De Schutter, UN Special Rapporteur on the Right to Food

Mr Chair, Excellences, Ladies and Gentlemen:

What I would really like to do today is to pay homage to the work of the Right to Food Unit of FAO.

Gabriele Zanolli / FAO

This Unit remains a minority voice in the broad debate on food. There exists a larger voice, which is vociferous at times. It sees food availability as the main problem, and increased food production as the solution. This is indeed the core business of FAO. It is the core business of agronomists and of economists whose work is to achieve the best, most efficient allocation of resources in a world of scarce resources and who are trained to produce more with less – not to distribute fairly.

The voice expressed by the right to food defenders is distinct. This minority voice tells us that food availability may be a problem at times – for instance following drought or floods, or in conflict situations where food must be brought in from food surplus areas to food deficient regions. But, they add, food availability is not the problem: it is one of a number of potential causes which may lead to hunger and malnutrition. The cause of hunger and malnutrition may indeed lie in discrimination, lack of accountability and social inequalities resulting in a situation where there are hungry and malnourished people although there is plenty of food available.

I should stress that these two views are not incompatible: there must be enough food for all before we can discuss questions of accessibility and equitable distribution of resources. But neither would it be absolutely right to say that the two views are complementary to one another because, in fact, they are not on the same plane. I believe that one of these, the minority view, has a richer diagnosis to propose. It is more lucid about the deep causes of hunger. It is also a voice that is more disquieting because it challenges the power of technocrats who see the question of hunger as a mere technical issue – deciding which seeds, and how much pesticides and fertilisers, are required to ensure that enough food is produced.

Instead, addressing the question of hunger and malnutrition from the point of view of the right to food poses the question of power: how power is distributed and how it is exercised. No wonder, then, that this minority voice is sometimes derided, ignored or even repressed. I have seen it myself first hand in my exchanges with governments and agencies on the responses to be given to the global food crisis.

Many want more food to be produced, but they forget to ask by whom and for the benefit of whom – as if more food would automatically alleviate the fate of the hungry. This is equivalent to saying that having more Wall-mart stores in New York would solve the problem of hunger in that city. They want to invest more in agriculture, and they are right to do so: reinvesting in agriculture, a neglected sector for so many years, is absolutely essential. But they forget to ask which kind of agriculture: agro-industrial agriculture? Or one that would sustainably keep smallholders in business? They want, and indeed we all would want, the prices to go down on international markets but they forget that for many years impoverished countrysides have subsidized the cities by dumping cheap food in urban centres at the expense of the livelihoods and, sometimes, the very survival of smallholders. They do not see that the real problem is not high prices but rather the insufficient purchasing power of the poor and the widening gap between farm gate prices and the prices paid by the consumer at the end of the food chain. They want more international trade. But they forget that all too often, trade – if not adequately regulated – has benefited only a privileged minority, and has increased inequalities and the dualization of the farming sector, further marginalizing family farming.

Is it not extraordinary that 60 years after the Universal Declaration of Human Rights, those who insist on the centrality of the right to food in the debate on food security, those who insist on food being more than a basic need to be fulfilled by public policies, a human right that requires accountability mechanisms for its effectiveness – are still a minority rebelling against the mainstream view?

The Right to Food Unit of FAO is the vanguard of a programme of action: the programme of the defenders of the right to food. And this is a programme that all of you in this room are part of today.

The programme has three components: the first is that of broadening and strengthening the remedies available for victims of violations of the right to food; the second is ensuring institutional mobilization beyond courts; and the third is developing the normative content of the right to food.

This programme is first about improving remedies. Significant progress has been made towards justiciability – exigibilidad – of the right to food, particularly before national courts, on the basis of the principles of non-discrimination, non-retrogression (understood as the prohibition to take steps backwards) and the judicial imposition of duties on public authorities, defined by national legislation. And indeed, one of the main advantages of a framework law is to define such duties in order to allow for judicial control: framework laws empower courts by making it possible for them to uphold the right to food without being accused of judicial law making – of 'legislating from the bench.'

This development toward the justiciability of the right to food shall be pursued further. I believe that the entry into force of the optional protocol to the International Covenant on Economic, Social and Cultural Rights shall impact this development very significantly, percolating down to national courts.

The second component of our programme is institutional mobilization beyond courts, not only because courts will effectively protect the right to food only if they have the support of a broader social movement – since this is a condition of the legitimacy of courts in the long term – but also because courts are not always well suited to protect the right to food, for three reasons: *first*, they need to receive claims, actions by victims who may face many obstacles, particularly in the absence of class action, from group action mechanisms; *secondly*, because courts may, at best, strike down, or not, legislation but they cannot create new laws when the regulatory framework is deficient; *thirdly*, courts intervene on an *ad hoc* basis, and therefore they generally cannot follow up on the remedies they prescribe nor monitor implementation over long periods of time. In situations where, for example, there is a need for agrarian reform, for improving the organization of farmers into cooperatives, or where marketing boards are to be re-established, courts are powerless to bring about such change. Although there are some exceptions, particularly from the Indian Supreme Court, these remain few and far between, and will not easily be replicated in other jurisdictions.

This is why the very promising development we are now seeing within the UN Committee on Economic, Social and Cultural Rights – with the Indicators, Benchmarking, Scoping and Assessment (IBSA) procedure now being road-tested within the committee, on the initiative of Eibe Riedel, vice-chair, and FIAN – cannot be replicated at national level by courts. This procedure is based on the definition of indicators and benchmarks, followed by a process of scoping, in dialogue between the Committee and governments, and finally by a regular assessment of progress made. Such a procedure, interesting and innovative as it is, requires a form of control which is spread over time: monitoring progress made at regular intervals – a task which a judicial body is usually ill-suited to perform. Therefore, institutions other than courts need to be involved in the realization of the right to food.

There has been much emphasis recently on the role of governments, the executive branch. We insist, for example, on inter-ministerial coordination, on support at the highest political level. These are amongst the 'lessons learned from Brazil' – to borrow from the title of a brochure prepared by the Right to Food Unit in 2007. But parliaments also have a role to play. Parliaments are not simply there to legislate, by voting on the laws presented to them for approval. In mature democracies, their role is increasingly that of controlling the executive by ensuring the participation of civil society organizations, to debate reforms, and request the government to explain the choices they make, thereby improving transparency and accountability. Indeed, it is on this theme that I shall focus my proposals to the Inter-Parliamentary Union meeting on the global food crisis in Geneva, in a few days' time.

National human rights institutions also have a tremendously important role to fulfil. They present five advantages over courts. *Firstly*, such institutions are proactive rather than simply reactive; they are proactive in that they do not depend on the vagaries of individual initiatives but can anticipate problems in order to propose solutions. *Secondly*, national human rights institutions or commissions have the ability to ensure the follow-up of their recommendations and can exercise pressure on governments to act upon such recommendations. *Thirdly*, they have greater flexibility in the remedies they can afford, both individually, for individual victims, and collectively, when the problems are of a more structural nature. *Fourthly*, national human rights institutions may more easily rely on States' international obligations which are contained in norms, which are not self-executing and which therefore courts themselves might be hesitant to take as grounds for their decisions. National human rights institutions may take into account international treaties or other sources of international human rights law, despite the lack of precision or clarity of the principles on which they rely. *Fifthly*, national human rights institutions are ideally placed to involve civil society organizations, and non-governmental organizations in monitoring the work of the executive branch of government.

Finally, there is a third component of the programme of action which defenders of the right to food have today for themselves. This is the development of the normative content of the right to food.

There are, I would suggest, five areas where the right to food requirements remain underdeveloped or difficult to monitor, and where we need to make further progress. The first such area is in the management of food aid: how to improve transparency and accountability in the way in which international food aid is being used and distributed. This is one of the main stakes in the renegotiation of the Food Aid Convention, currently being discussed. The second area which, I believe, deserves our attention, is the place of the right to food in the negotiation of international agreements on trade and investment. All too often these agreements are negotiated by the executive with little or no oversight on the part of parliaments and without taking into account the right to food. Parliaments are placed before the fait accompli when asked, finally, at the end of a long process of negotiation, to ratify whatever has been negotiated. As a minimum, right to food impact assessments should be performed on the draft proposals which are being submitted in such negotiations. A third area where more work needs to be done is in the preparation of public budgets. Here again, government is often the sole arbiter between competing priorities – education, health, agriculture, national defence – and parliaments generally defer to the judgement of the executive on this issue.

If the right to food is to be taken seriously, it requires obliging a government to justify its choices, taking into account the international obligations imposed on governments.

In these three areas, for a variety of reasons, governments are under very little scrutiny, if any, by national parliaments or civil society organizations. And the challenge, I think, is how to implement, in these fields – food aid, trade and investment, international agreements and the formation of public budgets – what has been referred to as the 'PANTHER' requirements, an acronym forged by the Right to Food Unit, referring to the values of participation, accountability, non-discrimination, transparency, human dignity, empowerment and the rule of law. Should we, for example, insist on food aid being distributed in accordance with the legislation describing how to map the needs of the hungry in order to ensure adequate targeting? Should we impose impact assessments regarding the risks to local agriculture producers, in the distribution of food aid? Should we force the idea that a predefined percentage of public budgets be earmarked for agriculture or, even more specifically, to support family farming? These are challenging questions which are posed to us in the three areas of food aid, trade and investment, and public budgets.

However, we also encounter the same kind of difficulty in two other areas. The fourth area is in controlling the role of international organizations, including – but not limited to – international financial institutions. Should we insist on the member states of these organizations exercising a due diligence control on how the said organizations operate? Do they comply with the right to food and should Member States be the guardians of how they do this? Or else, should direct obligations be imposed on such international obligations under general international law? Or again, should the obligations to protect and fulfill the right to food be imposed within their mandate? If we choose this second route, then how do we ensure participation, transparency and decision making within these organizations? How can we reconcile this with the principle of specialty of international organizations, the principle according to which they may only adopt measures that are within their mandates?

Finally, in a fifth and last area, the question is not only how the right to food can be implemented but also what it means, and which obligations it imposes. And this fifth area is the responsibility of private actors in implementing the right to food – providers of inputs to agriculture, food processors and traders, and food retailers. I believe that there is an urgent need to clarify what it means precisely for these actors to respect the right to food and, consequently, what measures the State should take in order to regulate the behaviour of the very influential and increasingly concentrated private actors in the food sector. I intend to convene a consultation in Berlin in June 2009, to examine this issue in greater detail.

I would like to close with this and thank you for your attention. I do look forward to our working together. Thank you.

3. Forum Orientation by Barbara Ekwall, Coordinator, Right to Food Unit, FAO

Background

Four years ago, we still did not know if there would be any Right to Food Guidelines or, to give them their full title: *Voluntary Guidelines to Support the Progressive Realization of the Right to Adequate Food in the Context of National Food Security*. The adoption of the Guidelines by FAO Council in 2004 is indeed a milestone in the development of the right to food. It reflects FAO's vision of a world without hunger, made possible by linking food security instruments with human rights and governance tools to tackle the root causes. In the same year, FAO Council recommended that FAO member countries implement the Right to Food Guidelines and asked the Secretariat to support them in their efforts. This presentation will look at developments with regard to implementation from FAO's perspective.

Gabriele Zanolli / FAO

Five Areas of Activity

The Right to Food Unit was created in 2006, with four main areas of activity: The first area concerns capacity development, and building awareness and understanding of the right to food. As the right to food is a new concept, this activity was essential and a pre-requisite for work in other areas. World Food Day 2007 constituted a major contribution towards this objective: the unprecedented mobilization worldwide on that occasion confirmed both relevance of the right to food and the interest it has brought to bear. Another major achievement has been the creation of a right to food website, with an average of 8 500 visitors per month. This website was placed on the Yahoo search engine initially and then on Google, using the keywords 'right to food.'

The second area of activity relates to the development of tools, methodologies and studies to support the implementation of the right to food. Detailed guidelines were developed on how to legislate, monitor, assess and budget right to food. These guidelines are available for the Forum participants.

The third area relates to integrating the right to food in FAO's work. Mainstreaming has been particularly successful in dealing with concrete projects and undertaking activities jointly with other departments or programmes, such as those undertaken together with FAO's Knowledge Exchange and Capacity Building Division, the Forestry Department, the Nutrition and Consumer Protection Division and the Special Programme on Food Security.

The fourth area and objective concerns providing support to countries based on their requests, which have varied from *ad hoc* assistance to specific processes, to more comprehensive projects covering several mutually reinforcing areas of right to food activities.

The Seven Implementation Steps

Our initial experiences show that the national implementation process evolves around seven practical steps, with capacity development as an integral part of all of them.

The first step is to identify who the hungry people are, where they live, and why their right to food is not being realized. In-depth knowledge of the hungry and of the underlying causes of their food insecurity are essential to enable governments to target policies, laws, institutions and budgets aimed at realizing the right to food. The need for disaggregated data cannot be over-emphasized. Such assessments have already been conducted in Bhutan and the Philippines.

As a second step, countries can assess policies, institutions and budget allocations to better identify both constraints and opportunities in realizing the right to food. This assessment will indicate what policy changes and new measures are required to improve food security for all, as a human right. Such an analysis has been undertaken in the Philippines and Mozambique.

Thirdly, food security strategies will build on the above assessment and causal analysis, and provide a roadmap for coordinated government action to progressively realize the right to food. This includes developing food and nutrition security strategies which should have targets, time frames, clearly allocated responsibilities and evaluation indicators that are known to all. Strategies will explore immediate relief measures, as well as the creation of a conducive environment that enables every person to feed himself or herself by their own means. Some countries have already developed food and nutrition security strategies that focus on the right to food, among them Tanzania, Kenya and Mozambique.

The fourth step concerns the roles and responsibilities of different government sectors and levels which need to be clearly defined and communicated to ensure transparency, accountability and effective coordination. This is an essential step for the implementation of strategies, policies and programmes.

An important fifth step is achieved when the right to food is integrated into legislation, such as a constitution or framework law, thus setting a long-term binding standard for government and stakeholders. Several examples will be discussed in this Forum, such as those of Brazil, Bolivia and Guatemala and, most recently, the approval of the Constitution of Ecuador.

The sixth step concerns monitoring. Monitoring the impact and outcomes of domestic policies, programmes or projects will make it possible to measure the achievements of stated objectives, fill possible gaps and constantly improve government action.

Finally, the implementation of the right to food necessitates the putting in place of recourse mechanisms to enable right holders to hold government accountable – the seventh step. A right is not a right if it cannot be claimed. Such mechanisms can be judicial, involving an action in court, or extrajudicial (ombudsperson, human rights commission). It is essential to incorporate operational or administrative recourse mechanisms at project or programme level, to ensure that corrective measures are taken without delay – for example, in the context of delivery of services such as social safety nets or school feeding programmes.

Progress in implementation during the short period that has elapsed since the adoption of the Right to Food Guidelines clearly indicates that, for many countries, the right to food is here to stay.

What is this Forum about?

The Forum will demonstrate, with practical examples, how the right to food can contribute to promoting food security for all. It is about sharing experiences and learning. Up to now, several initiatives have been taken by different stakeholders to promote the right to food or certain aspects of it. Valuable experience has been gained and progress achieved, mostly in the context of 'pilot projects'. This Forum is the first ever platform where the lessons learned and individual country experiences can be exchanged, tested and validated among stakeholders at an international level. Exchange of this type is extremely important for an emerging issue such as implementation of the right to food. It helps to identify what areas need to be strengthened and affirms which choices were the successful ones. It also provides new insights and ideas to be pursued in the future.

The Forum is a platform for multi-stakeholder dialogue. During the negotiations on the Right to Food Guidelines, civil society organizations and other stakeholders played an important role. They continue to be an important motor for the right to food agenda and a valuable partner in many countries, supporting government action on the right to food. The Forum aims to strengthen this partnership.

The Forum is about knowledge. Knowledge is a resource, a global public good, which is not depleted through use. On the contrary, the more knowledge is shared, the more powerful it becomes. The more knowledge is confronted with other insights, the more it is developed, enriched and made relevant for practical use.

Most importantly, the Forum is about strengthening and further promoting the implementation of the right to food. It is not an end – rather, it is the beginning of a new phase of implementation, with greater focus on country level activities, using the knowledge, tools, networks and strategies developed up to now.

Excellencies, Ladies and Gentlemen, dear colleagues and friends: This is your Forum. I wish you many fruitful exchanges, enriching discussions, stimulating networking and strengthened commitment to promote the realization of the right to food.

II. FINAL REPORT BY MARC COHEN, FORUM RAPPORTEUR

In this report, Marc Cohen summarises many of the important contributions made by participants during general plenary discussions and in the thematic panel sessions He also provides some general reflections as Forum Rapporteur.

Gabriele Zanolli / FAO

The context for the Forum was shaped by the current food crisis due to soaring food prices – a crisis that has wiped out four decades of progress against hunger. There is serious concern that the prevalence of hunger in the world has increased, perhaps even significantly, and will continue to do so in the coming years. Thus, the right to food is now a matter of urgency, and it is essential that we understand why this is so.

Right to Food Approach

One may ask: what is the added value of the right to adequate food? How can right to food approaches contribute to solving or mitigating the food crisis? Is it not sufficient for governments to invest in agriculture, rural development, rural and urban food security and nutrition? An important added value of the right to food lies in the realm of governance: good governance involves empowering people to be active participants in decision making, and in the creation of recourse and accountability mechanisms. The right to food also provides new insights into the causes of food insecurity, beyond inadequate food availability or low incomes – such as discrimination and socio-economic exclusion. Everyone has a right to food – and like all human rights, the right to food is universal. The participants stressed the fact that the right to food approach to food security is based on giving the highest priority to those who suffer hunger, are food-insecure or vulnerable to food insecurity. This approach sees people as actors in achieving food security, and not as mere objects of development policy. The almost one billion hungry people are no longer the problem; rather, they form a key part of the solution. If the approach to achieving food security is limited to hoping that governments will do the right thing without pressure from the people who should hold government accountable for progress, we may have a very long wait before hunger is eliminated.

Forum participants representing governments, civil society, international organizations and academia all voiced strong support for the work of FAO's Right to Food Unit. Attention was drawn, in particular, to the wealth of studies, country reports and methodological tools that the Unit has produced, including a right to food curriculum. Most of these documents are accessible via multi-media and include practical manuals and instruments that can be used by FAO Member States, civil society and others. The Right to Food Unit has continually stressed the key role of what it calls the PANTHER principles – participation, accountability, non-discrimination, transparency, human dignity, empowerment and rule of law – as the foundation for realising the right to food.

Participants felt the work of the Unit should continue and could focus, for example, on facilitating a network of educational and training institutions that deal with the right to food, as one of its important tasks. There is also a valuable knowledge management role for the Unit to fill.

Participants urged each other to inform their countries' Permanent Representatives to FAO of the Unit's valuable work towards the implementation of the right to food at country level, as well as in FAO's own work.

Many participants echoed the words of Martin Nissen from the German Embassy in Paris, who played a significant role during the formulation of the Right to Food Guidelines: *'The right to food is not about obscure theory or highly technical procedures: it is about practical and effective solutions and actions.'*

Participatory Processes

A recurrent theme at the Forum was the important role of civil society in encouraging government action, facilitating the empowerment of vulnerable people and fostering accountability. Many right to food NGOs have focused on court actions or strengthening national human rights institutions but they also have a key role to play in other areas. Both government and civil society representatives at the Forum stressed the need for participatory processes and broad multi-stakeholder consultations in the development of laws and institutions to implement the right to food.

There were significant exchanges on country level experiences in implementing the right to food and many challenges were identified. It was noted, in particular, that mainstreaming the right to food in the formulation of national strategies, policies and plans at the country and global level is still 'work in progress': too often the right to food is excluded from poverty reduction strategies. Discussions also centred on how trade, investment and agricultural production relate to the right to food. Additional challenges identified included the need for inter-institutional coordination, policy coherence and aid effectiveness.

National Leadership

Strong national leadership is extremely important in promoting the right to food. There were several examples from Brazil, Mozambique and Guatemala, where presidential leadership on the right to food has made a significant difference. Parliaments and national human rights institutions also have an important role to play. The role of parliaments goes well beyond the important job of passing laws: it includes oversight of the executive branch, facilitation of popular participation in policy making and implementation, and promotion of government accountability. National human rights institutions should be independent (in accordance with the Paris Principles) and should be able to initiate investigations on their own, serve as quasi-judicial bodies and make recommendations to the government on remedial actions.

Laws, policies and programmes are essential for the realization of the right to food. Their implementation is of crucial importance and must be carefully monitored to ensure full compliance with human rights principles. National budgets also need to be analysed and monitored to see the extent to which they reflect right to food priorities and provide adequate support in the implementation of right to food measures and actions.

Learning from Country Experiences

The five case studies presented featured concrete country experiences[239], providing an opportunity to learn from best practices and understand what did not work well and why. Both policy makers and civil society leaders contributed additional country level information. An example from Brazil was the Fome Zero (Zero Hunger) Strategy, which incorporates a right to food perspective and has contributed to substantial poverty reduction. The government devotes more than US$ 6 billion a year to one of the key elements of the Strategy, the Bolsa Familia Programme, which puts cash into the hands of poor families.

Political will on the part of government can make a real difference in the context of a strong civil society anti-hunger movement. We learned of the important role of the media in Guatemala, and of the need to focus on capacity building at the local government level. The well-known Supreme Court decision in India on the right to food has extended public food programmes to millions of people. India's right to food movement also succeeded in getting the national employment guarantee enacted – a real milestone for the country. It has become clear that implementation and follow-up of a court decision requires considerable data collection and monitoring. There have been locally owned and driven efforts in both Mozambique and Uganda to incorporate the right to food into framework law, and its use to enhance existing food security and nutrition strategies and policies. This resulted from broad, multi-stakeholder processes. The creative use of the Right to Food Guidelines in the Philippines was also highlighted. This involved undertaking a comprehensive assessment of national laws and institutions, surveys of vulnerable groups, and the application of locally adapted indicators in monitoring the right to food, including their insertion in community monitoring systems.

Emerging Jurisprudence

With regard to access to justice and legislation, right to food jurisprudence is beginning to emerge. Cases were cited from South Africa, Switzerland, Nepal, India and, going back to the 1960s and 1970s, the United States of America. Recourse mechanisms differ according to country contexts and the nature of legal systems, and embrace judicial, quasi-judicial and administrative bodies. The India example shows that the ability to bring public interest litigation and the availability of a public interest bar are important.[240] This is now also the case in Nepal, but such mechanisms are not available everywhere. Right to food framework laws are being adopted, or are under consideration in Brazil, Guatemala, Nicaragua, Peru, Mozambique and Uganda. Ecuador has just adopted a new constitution in which it is explicitly stated that the right to food is justiciable. In the case of countries that have ratified international treaties with right to food provisions, the right to food is then automatically incorporated into national law.

Participants recognized the importance of working at the sub-national level, including at the district level; local governments are increasingly responsible for policy implementation and therefore need to be engaged on the right to food issues. Local capacity needs to be developed and strengthened for both governments and local institutions. The latter should be enabled to monitor the implementation of policy measures and public service delivery.

---

239 The case studies referred to, from Brazil, Guatemala, India, Mozambique and Uganda, are provided in Part III of this document.

240 India's Colin Gonsalves, the quintessential public interest lawyer, was among the Forum participants.

In his opening keynote presentation, Olivier De Schutter, United Nations Special Rapporteur on the Right to Food, raised some provocative questions regarding the obligations of international organizations, particularly the international financial institutions and the WTO, and of non-state actors. The right to food perspective can play a very important role in some upcoming international negotiations, particularly those on climate change and the renewal of the Food Aid Convention. On the last day of the Forum, there was further discussion on the need for a global right to food strategy and the question of food sovereignty.

Capacity Building for All

In the session on capacity development it was pointed out that both right holders and duty bearers have considerable capacity building needs in relation to the right to food. For example, people in the North need to be sensitized to the rights-based approach to development; right holders need to understand their rights and the processes to be undertaken in order to bring claims and they also need to be aware of how to hold officials accountable. Duty bearers, including lawyers, judges and civil servants, need to be trained with regard to the implementation of their respective right to food obligations. University training on the right to food should be offered not only to regular students but also to government officials and representatives of civil society organizations. Right to food training should be demand-driven, for example, by carrying out a careful capacity gap analysis. It is important that academic institutions maintain their independence, even when they engage with governments in providing right to food training. In Brazil, distance learning has been successfully adopted to develop capacity among a wide range of stakeholders. Both the Right to Food Unit and the UN Office of the High Commissioner for Human Rights have played an important role in capacity strengthening at country level.

Information and Assessment

With regard to information and assessment, it was pointed out that participatory processes and high level consultations are of particular importance. Legal, institutional and policy frameworks need to be assessed through a right to food lens. This adds new dimensions to more traditional food security assessments. In this context, the Right to Food Unit has developed an assessment manual that has been used in the Philippines, Mozambique and Bhutan. As right to food assessment is a very new area of work, knowledge on how to conduct such assessments is currently being generated.

Effective Monitoring is a 'Must'

Effective monitoring is essential to determine whether progress is being made and whether governments are meeting their obligations. For this, disaggregated data (reflecting gender differences, urban/rural differences, indigenous/non-indigenous status, etc.) are indispensable. The generation and analysis of such data may require capacity development and indicators may need to be tailored to local needs. The Right to Food Unit has developed a comprehensive manual on monitoring. FIAN and Mannheim University are also developing right to food indicators through the IBSA/Indicators, Benchmarks, Scoping, Assessment project. Indicators must be simple, but not simplistic. Baseline data are important in establishing benchmarks and targets against which to monitor progress. In each case, there should be a clear indication of who is monitoring, what is being monitored and for what purpose. Right to food monitoring is a government responsibility but it can be done in partnership with civil society. Ideally, it should be fully integrated in other monitoring activities, such as poverty monitoring and programme monitoring. Monitoring by civil society organizations and academia may be undertaken as part of holding governments accountable, and to obtain independent information. For example, the so-called shadow reports prepared by non-governmental organizations and presented to the UN Committee on Economic, Social, and Cultural Rights, together with the government's report, serve such a purpose. The Committee, having ratified the International Covenant on Economic, Social and Cultural Rights, oversees States Parties' compliance with their ESCR obligations. There are many institutional and professional barriers for academia in undertaking the interdisciplinary work needed to contribute to the right to food field.

In right to food assessment and monitoring, qualitative approaches offer important insights that are not available from statistics, and should thus complement quantitative approaches.

Tailor-made Strategies to Fit Context

Participants emphasized that strategies for realising the right to food must be tailored to specific national circumstances and opportunities. For example, a legal strategy may make sense in a particular context, whereas a focus on political and social advocacy may offer a more viable approach elsewhere. Budget analysis and citizen audits are important tools in efforts to hold governments accountable for the implementation of the right to food. Here again, adequate capacity and resources are required. The realization of the right to food involves economic aspects through the combination of an enabling environment that: (i) expands people's livelihood opportunities, (ii) includes laws and policies that ensure vulnerable people's access to resources, and (iii) offers programmes that boost agricultural productivity and targeted safety nets.

There is a political aspect as well that encompasses democracy, equality, dignity and citizenship. Participants stressed the fact that strategies cannot focus on the national or local level alone, although they recognized that those are the crucial arenas for the right to food. However, global trade rules impinge on farmers' rights to save and re-use seeds and on national agricultural strategies. Many developing countries still have high levels of external debt. Unregulated transnational corporate activities may undermine the right to food. These issues are all directly related to the demand for food sovereignty advocated by many civil society organizations and some governments. Participants also emphasized that international institutions have obligations vis-à-vis the right to food. While the activities contributing to climate change are concentrated in the developed countries, the negative consequences for food security and the right to food are mainly felt in the developing world. 'Do no harm' is the key policy principle at both national and global level.

Policy Coordination

Policy coordination represents a substantial challenge, in light of the multisectoral nature of the right to food, and of food and nutrition security. Implementation of the right to food requires work across sectors, across various levels of government from the national to the local, and across lines of government, the private sector and civil society. A coordinating body within the national government can be helpful, but a broad sense of ownership on the part of all stakeholders in realising the right to food is likewise important. Harmonized efforts by the UN and donor agencies, in keeping with the One UN approach and the *Paris Declaration on Aid Effectiveness*, can play a crucially supportive role. A key aspect of coordination is the need to avoid duplication, to capture synergies and move beyond traditional sector-focused priorities and approaches.

The Way Forward

During the final plenary session, the Forum addressed the question of the way forward. Participants pointed out that there have been a number of right to food success stories in recent years, even though at a global level the number of hungry people has increased since the World Food Summit in 1996, reaching 923 million in 2009. The successes achieved offer opportunities for cross-country learning that can lead to sustainable progress. A right to food approach to food security can provide important support to efforts geared towards achieving the first Millennium Development Goal on cutting poverty and hunger. Governance issues, including the right to food and human rights principles more generally, must be considered in addressing the global food crisis. This also means taking into consideration the rights of smallholder farmers and other rural poor people. A focus on increased food production and safety net programmes is necessary but not sufficient. More attention should be paid to the right to food in emergencies and the fact that newly emerging issues, such as soaring food prices, bio-fuels, genetically modified organisms, speculative activities, seed patents and climate change, will all impinge on efforts to realize the right to food. These factors aggravate structural problems such as concentrated ownership of land, evictions, marginalization and exclusion, and urbanisation into slums. Poverty continues to co-exist alongside economic growth. United Nations bodies such as FAO play a crucial role in raising awareness with regard to alternatives but there needs to be greater collaboration across the UN system.

Key Actors – the Role of FAO

Participants identified the key actors engaged in advancing the realization of the right to food, as follows: individuals, non-governmental organizations, the Committee on Economic, Social and Cultural Rights, FAO, national human rights institutions, and this Forum. It was suggested, however, that there be a broader network for advocacy and communications on the right to food, and that communications be put into clear language that is accessible to policy makers and other non-specialists. It was recognized that whereas international strategies are important, states have the primary responsibility for realising the right to food within their territories. This task requires capacity, resources, government action and empowerment of the food insecure and vulnerable. Judicial and quasi-judicial bodies can play an extremely important role here. Greater attention needs to be given to the impact of sector policies on the right to food. Many participants underlined the important role played by FAO in supporting Member States' efforts to realize the right to food – particularly in recent years – in areas such as the drafting of legislation and the assessment, through a right to food lens, of existing national policies. It was pointed out that FAO can help sensitize governments to the opportunity offered by the right to food approach as a means to accelerate progress on food security. There are both technical and moral aspects to this. Participants encouraged FAO to make right to food one of the Organization's strategic goals, in conjunction with the reform process, and urged donors to provide adequate funding to FAO to enable it to continue its right to food activities. It was suggested that the Spanish Government's Millennium Development Goal Action Fund might be helpful in this regard.

Two concluding points: First, in addressing country-level issues, it is important not to engage in a 'one size fits all' approach: it is necessary to adapt right to food tools as required, to fit the specific country context. Second, those of us who have worked in FAO for a long time – as staff, representatives or advocates – will have noticed that we have come a long way since the mid-1990s, when the right to food was little known beyond the concerns of a few technical experts. As Barbara Ekwall, head of the Right to Food Unit, pointed out in her remarks on the first day of the Forum, *"The right to food is here to stay"*.

" The Right to Food is here to stay "